THE LITTLE ENGINEER COLORING BOOK

NUMBERS

SETH MCKAY

The Little Engineer Coloring Book: Numbers by Seth McKay
www.TheLittleEngineerBooks.com

Copyright © 2019 by Seth McKay

All rights reserved. No portion of this book may be reproduced, stored in a retrieval system, or transmitted in any form or by any means electronic, mechanical, photocopy, recording, scanning, or other-except for brief quotations in critical reviews or articles, without the prior written permission of the publisher.

Creative Ideas Publishing titles may be purchased in bulk for educational, business, fund-raising, or sales promotional use. For information, please email permissions@TheLittleEngineerBooks.com.

ISBN-13: 978-1-952016-01-1

Published by: Creative Ideas Publishing

Table of Contents

Introduction (*for Parents*) v

Introduction (*for the Little Engineers*) vi

1: There is 1 Earth 1

2: A motorcycle has 2 wheels 2

3: Water can come in 3 forms 3

4: A monster truck has 4 huge tires 4

5: The Saturn V has 5 powerful rocket engines 5

6: The crane has 6 wheels on each side 6

7: The helicopter has 7 windows 7

8: There are 8 planets are in our solar system 8

9: The transport truck is carrying 9 cars 9

10: The off-road vehicle has 10 lights 10

11: The tractor is pulling 11 bales of hay 11

12: The bucket wheel excavator has 12 bucket 12

13: There are 13 windmills in the field 13

14: The ferris wheel has 14 chairs 14

15: The dump truck unloaded 15 rocks 15

16: The space station has 16 solar panels 16

17: The cargo ship is carrying 17 containers 17

18: The train is pulling 18 railroad cars 18

19: The building will be 19 stories high 19

20: Cars can be 20 feet long 20

60: Cars drive about 60 MPH on the highway 21

206: There are 206 bones in the human body 22

767: Sound travels at 767 MPH 23

24,790: The fastest humans have traveled 24,790 MPH 24

29,029: Mount Everest is 29,029 feet tall 25

160,000: The fastest space probe went 160,000 MPH 26

670,616,629: Light travels at 670,616,629 MPH 27

Double the Fun! Pages repeated for extra number practice 28

Certificate of Completion 56

BONUS: Special Preview of Cars and Trucks Coloring Book 57

Introduction *(for parents)*

Thank you so much for your interest in The Little Engineer coloring books.

In this book your child will discover numbers. I thought it would be fun to teach numbers using awesome machines like rockets and excavators. Numbers 1-20 are introduced to help your child learn the sequence of numbers, but then I added in some fun, big numbers.

Some tips I recommend for using this book:

- First and foremost, have fun with your child as he or she is learning numbers and coloring!

- Sit with your child as he or she is coloring. Point to the number at the top of the page. Say the number out loud to your child and ask your child to repeat the number back to you. Afterwards, read the interesting fact at the bottom of the page. Finally, count the numbered objects on the picture by pointing to the numbers.

- It is best to color the pages in numerical order so your child begins to understand the order of numbers, but it is fine if your child would rather color in random order.

- When your child completes the coloring book, use the coloring book as a reading book. Read each page together and ask your child to point and count the numbered objects on each page.

Most importantly, just let your little engineer have fun!

Introduction *(for the Little Engineers)*

Hello Little Engineer!

My name is Seth, and I'm the Chief Engineer of this book. We will have so much fun learning about numbers and counting.

We will start by learning how to count 1-20 by counting fun machines, parts of cars, planets, and more. Then I will show you some cool big numbers!

Enjoy coloring, ask your parents a lot of questions, and most of all have fun!

1

0 1 2 3 4 5 6 7 8 9 10 11 12 13 14 15 16 17 18 19 20

THERE IS 1 EARTH.

EARTH IS THE PLANET WE LIVE ON, AND EARTH IS THE BEST PLACE TO LIVE IN OUR SOLAR SYSTEM.

The Little Engineer Coloring Book: Numbers

2

◎ 1 2 3 4 5 6 7 8 9 10 11 12 13 14 15 16 17 18 19 20

A MOTORCYCLE HAS 2 WHEELS.

BICYCLES ALSO HAVE 2 WHEELS, BUT TRICYCLES HAVE 3 WHEELS.

3

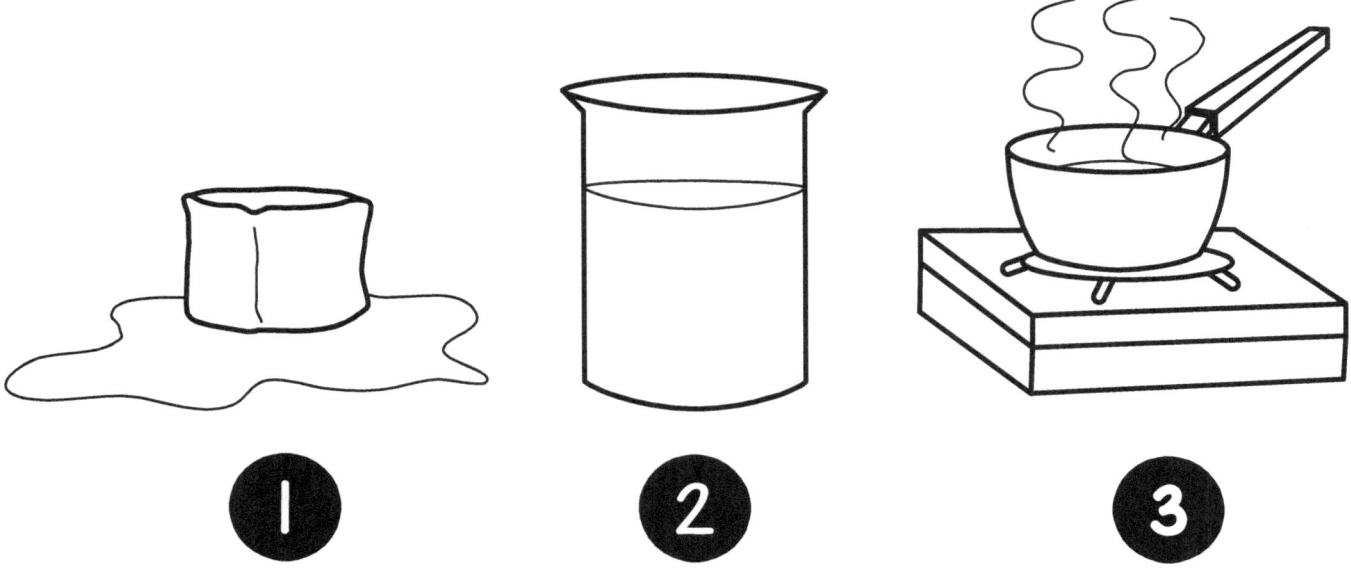

0 1 2 3 4 5 6 7 8 9 10 11 12 13 14 15 16 17 18 19 20

WATER CAN COME IN 3 FORMS.

WATER CAN BE SOLID (LIKE ICE CUBES),
LIQUID (LIKE THE WATER WE DRINK) OR
GAS (LIKE STEAM WHEN COOKING SPAGHETTI).

4

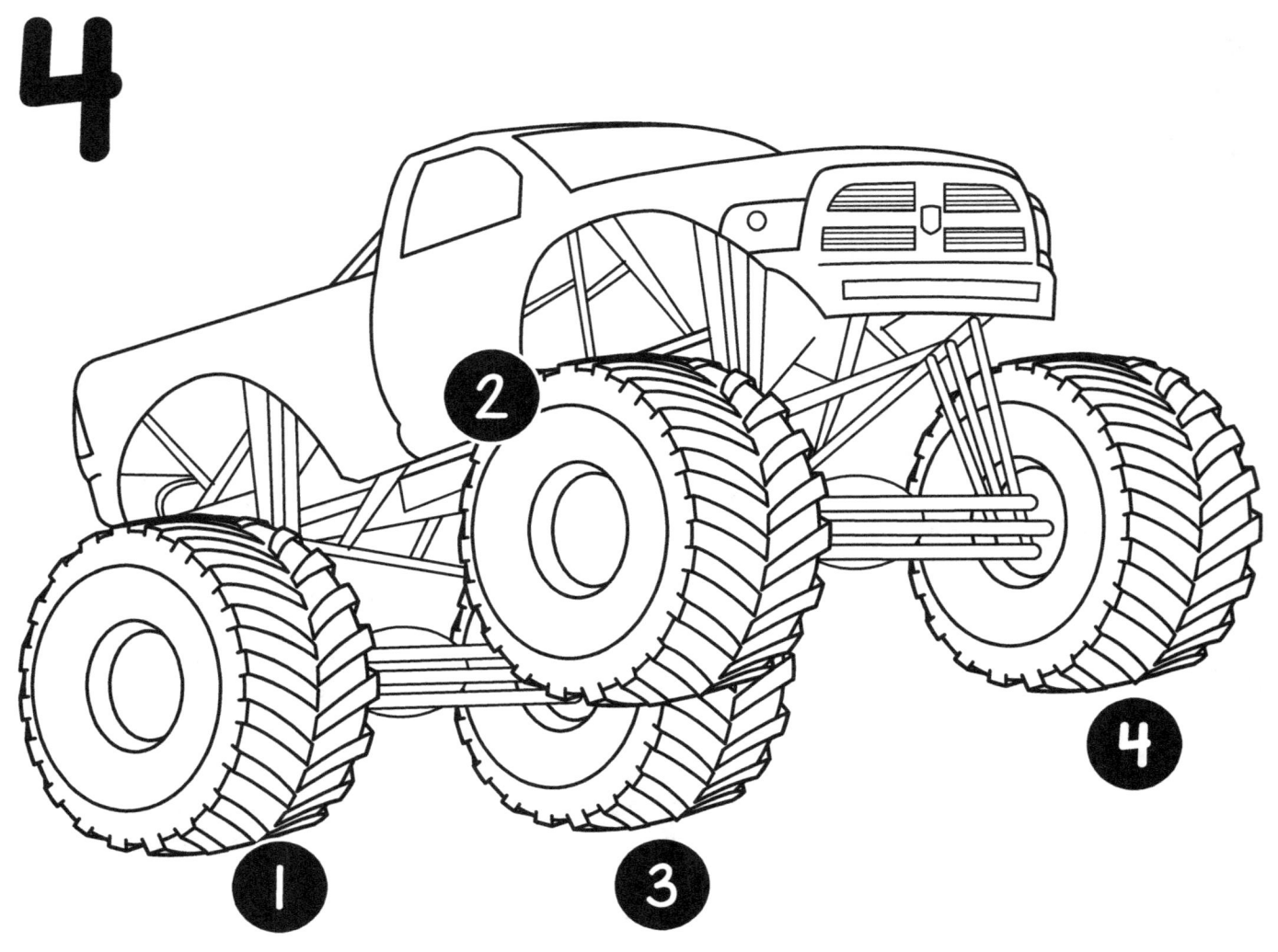

◎ 1 2 3 4 5 6 7 8 9 10 11 12 13 14 15 16 17 18 19 20

A MONSTER TRUCK HAS 4 HUGE TIRES.

THESE LARGE TIRES HELP THE TRUCK RUN OVER CARS!

5

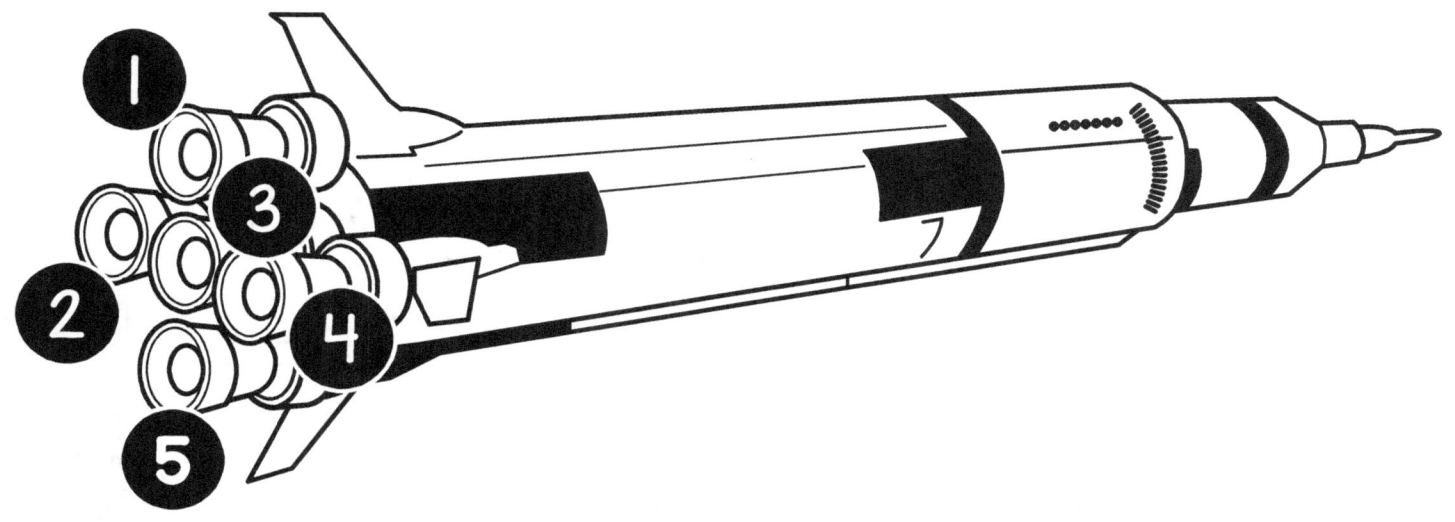

0 1 2 3 4 5 6 7 8 9 10 11 12 13 14 15 16 17 18 19 20

THE SATURN V (5) HAS 5 POWERFUL ROCKET ENGINES.

THE SATURN V TOOK ASTRONAUTS TO THE MOON 50 YEARS AGO!

The Little Engineer Coloring Book: Numbers

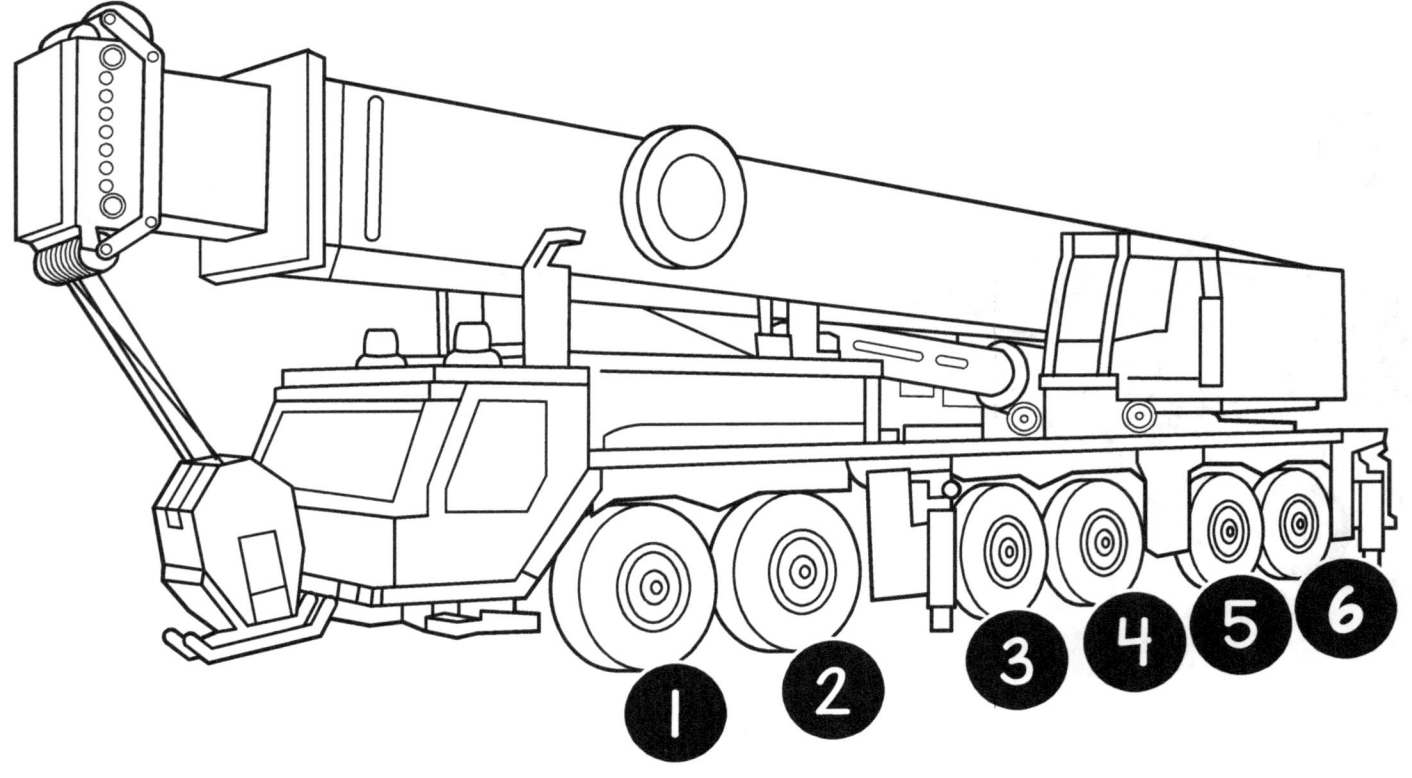

0 1 2 3 4 5 6 7 8 9 10 11 12 13 14 15 16 17 18 19 20

THE CRANE HAS 6 WHEELS ON EACH SIDE.

THIS MEGA CRANE IS INCREDIBLY POWERFUL, AND IS USED TO BUILD LARGE BUILDINGS AND BRIDGES.

7

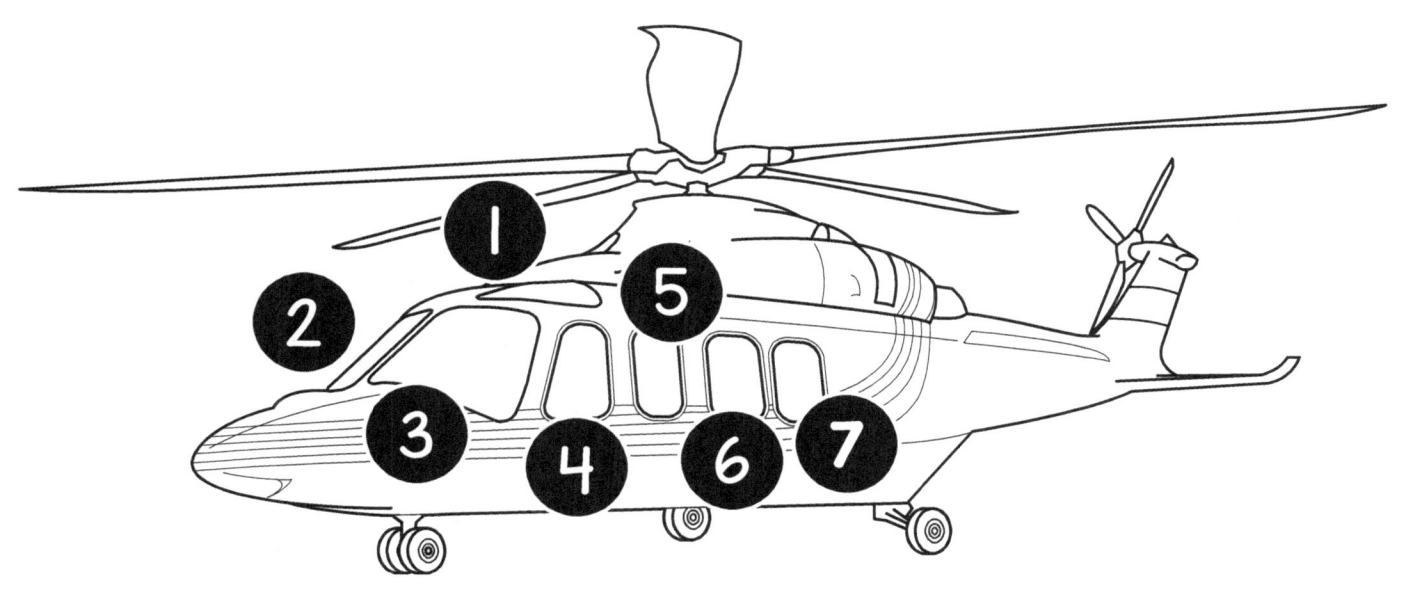

◎ 1 2 3 4 5 6 7 8 9 10 11 12 13 14 15 16 17 18 19 20

THE HELICOPTER HAS 7 WINDOWS.

HELICOPTER BLADES ACT LIKE A BIG FAN AND BLOW AIR DOWN WHICH MAKES THE HELICOPTER RAISE UP IN THE SKY.

The Little Engineer Coloring Book: Numbers

8

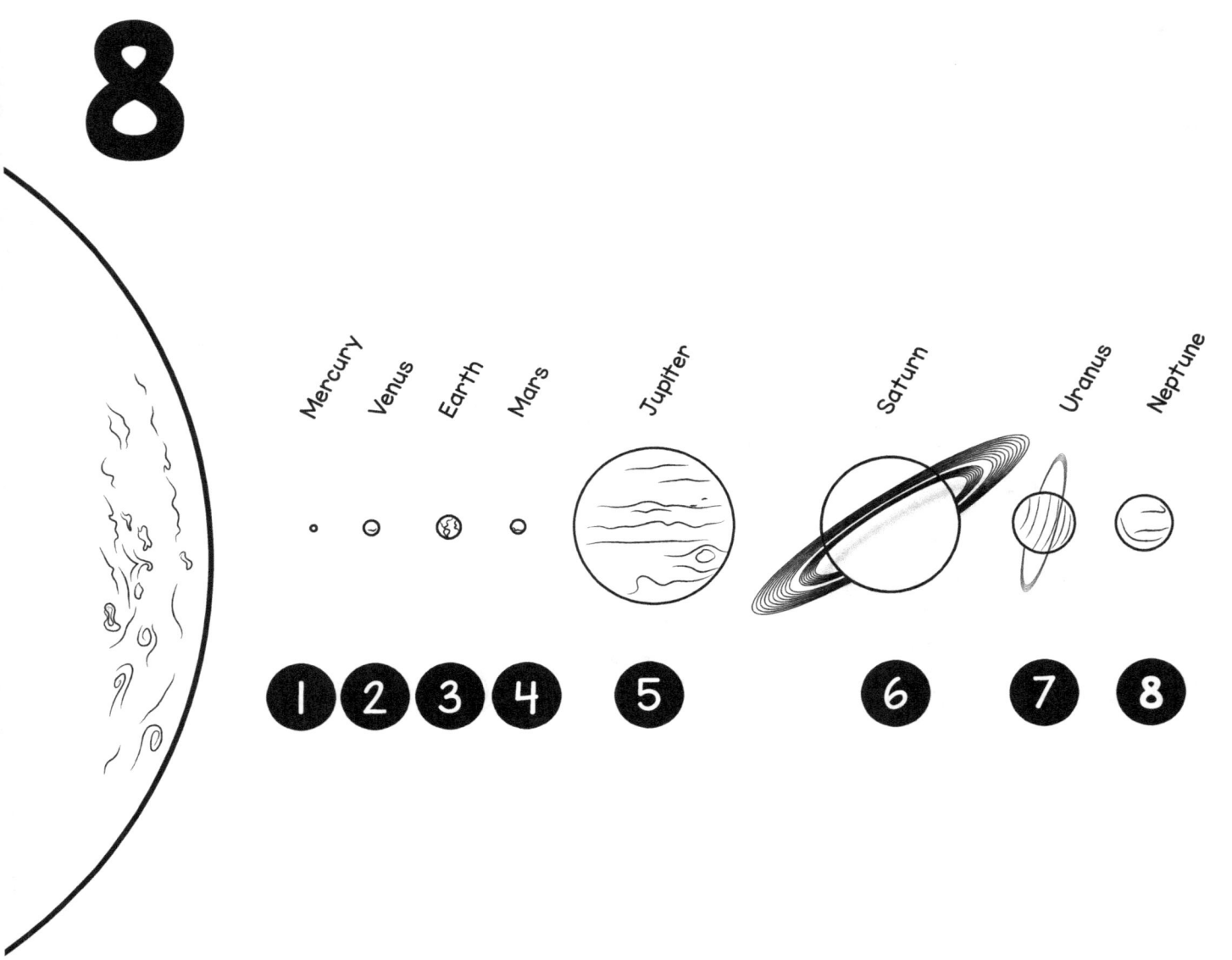

THERE ARE 8 PLANETS IN OUR SOLAR SYSTEM.

ALL OF THE PLANETS ORBIT (OR GO AROUND) THE SUN. THE PLANETS ARE FARTHER APART THAN SHOWN HERE.

9

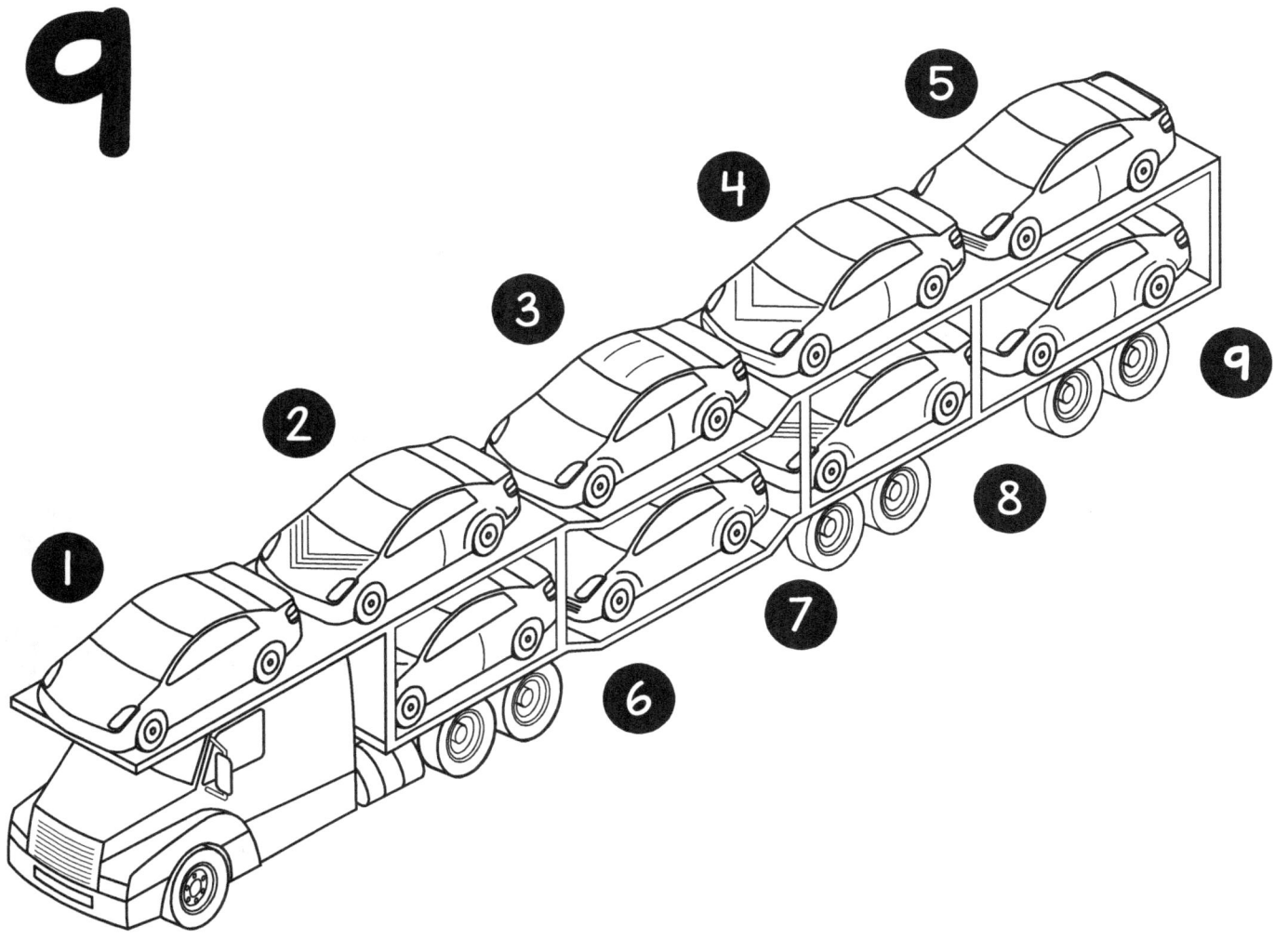

◎ 1 2 3 4 5 6 7 8 9 10 11 12 13 14 15 16 17 18 19 20

THE TRANSPORT TRUCK IS CARRYING 9 CARS.

BIG DIESEL TRUCKS ARE VERY POWERFUL, AND CAN CARRY MANY CARS.

10

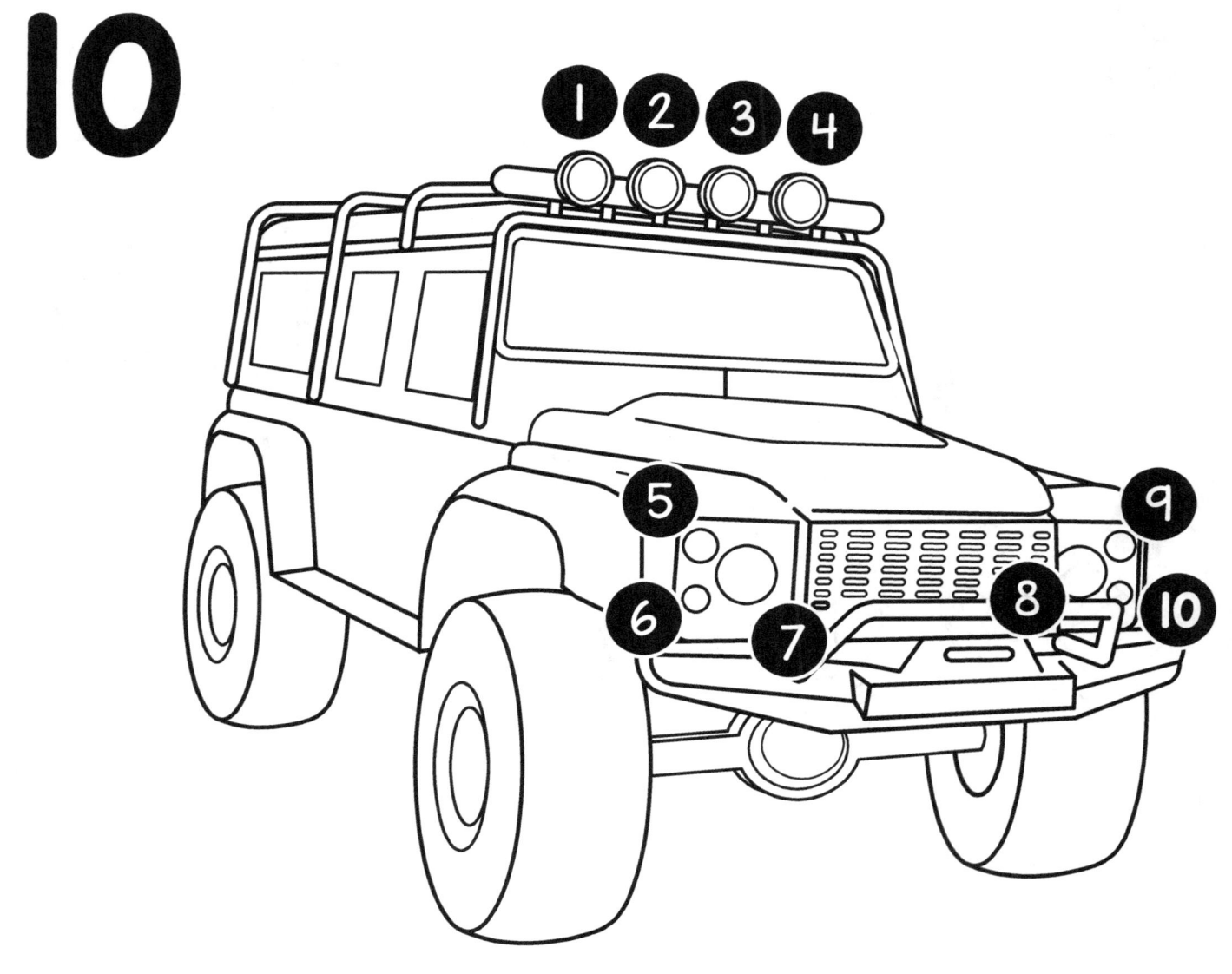

0 1 2 3 4 5 6 7 8 9 10 11 12 13 14 15 16 17 18 19 20

THE OFF-ROAD VEHICLE HAS 10 LIGHTS.

THE LIGHTS HELP MAKE SURE THE DRIVER CAN SEE WHEN DRIVING AT NIGHT.

0 1 2 3 4 5 6 7 8 9 10 11 12 13 14 15 16 17 18 19 20

THE TRACTOR IS PULLING 11 BALES OF HAY.

HAY BALES ARE MADE FROM CUT GRASS FROM THE FIELD, AND THEY ARE USED TO FEED THE COWS.

12

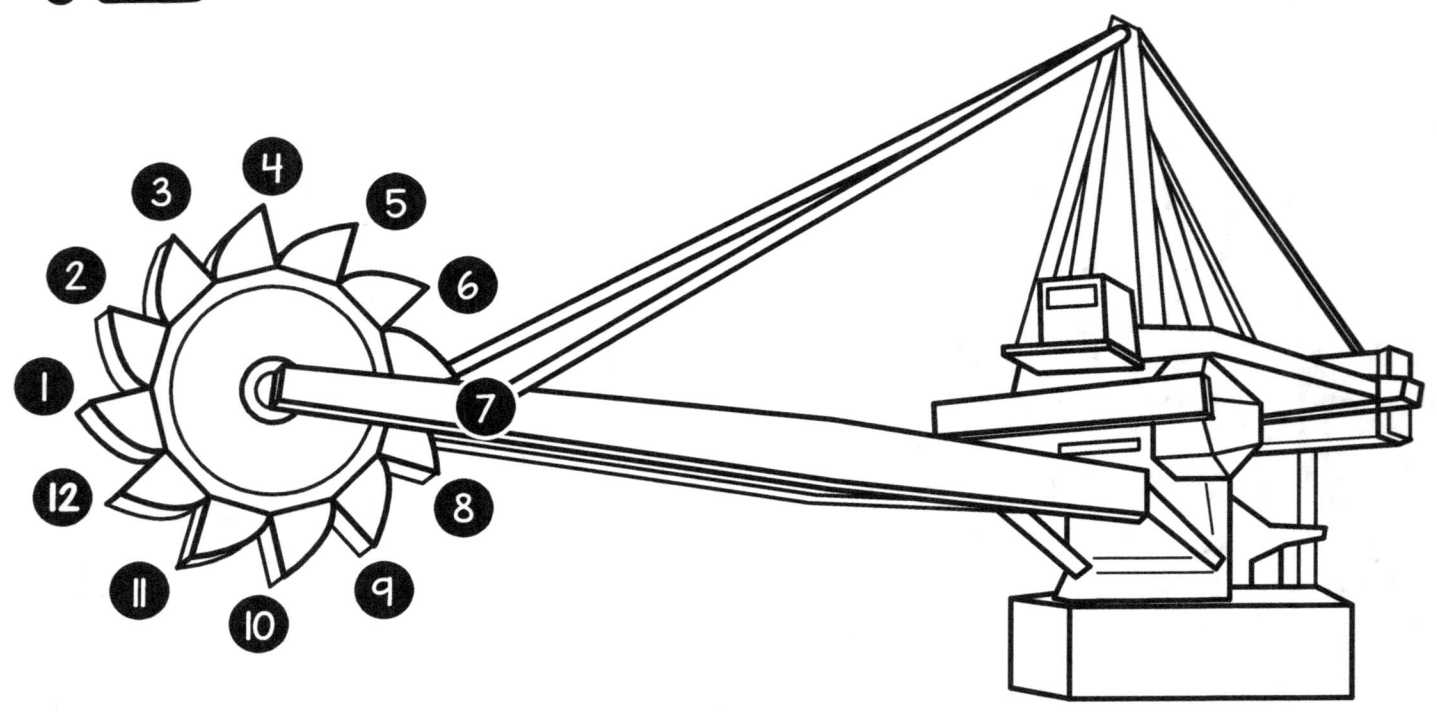

0 1 2 3 4 5 6 7 8 9 10 11 12 13 14 15 16 17 18 19 20

THE BUCKET WHEEL EXCAVATOR HAS 12 BUCKETS.

THIS EXCAVATOR IS BIGGER THAN YOUR HOUSE!
A CAR COULD FIT INSIDE EACH BUCKET!

13

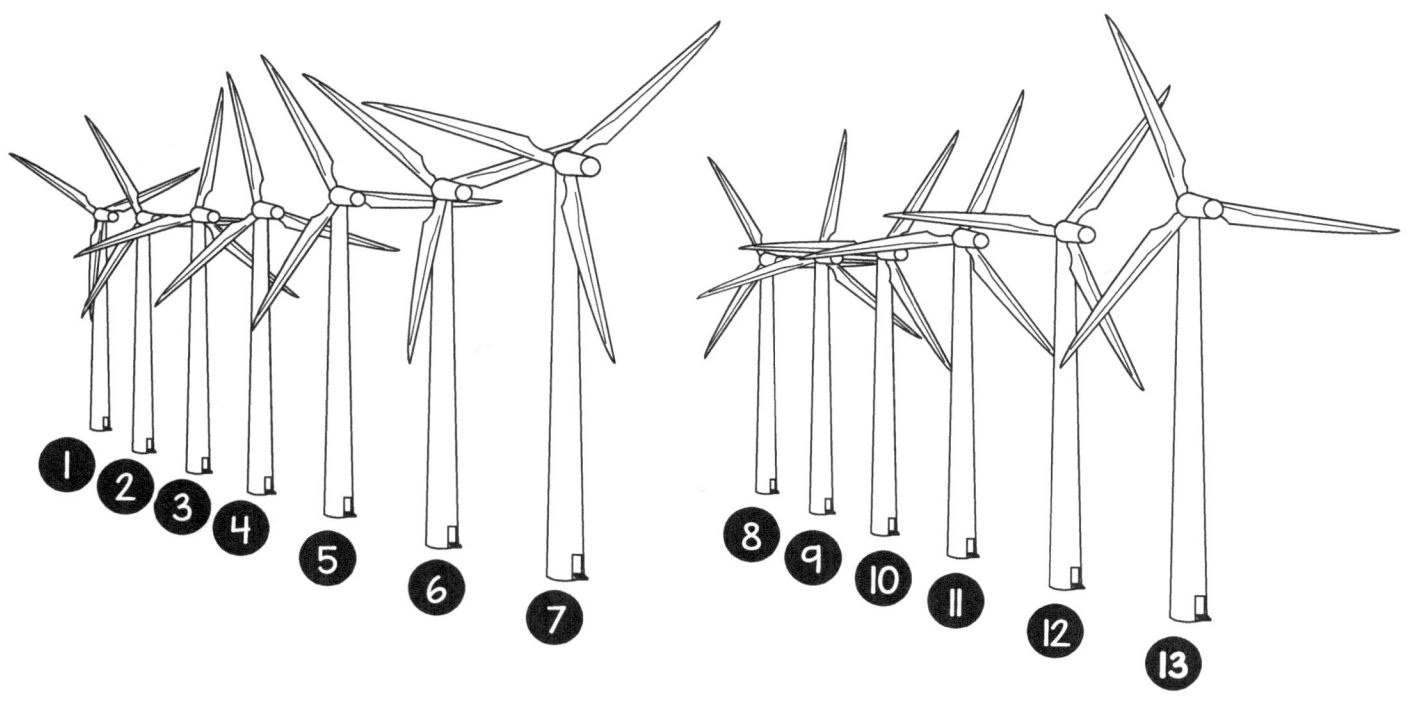

◎ 1 2 3 4 5 6 7 8 9 10 11 12 13 14 15 16 17 18 19 20

THERE ARE 13 WINDMILLS IN THE FIELD.

WINDMILLS GENERATE ELECTRICITY WHEN STRONG WINDS SPIN THE BLADES.

The Little Engineer Coloring Book: Numbers

14

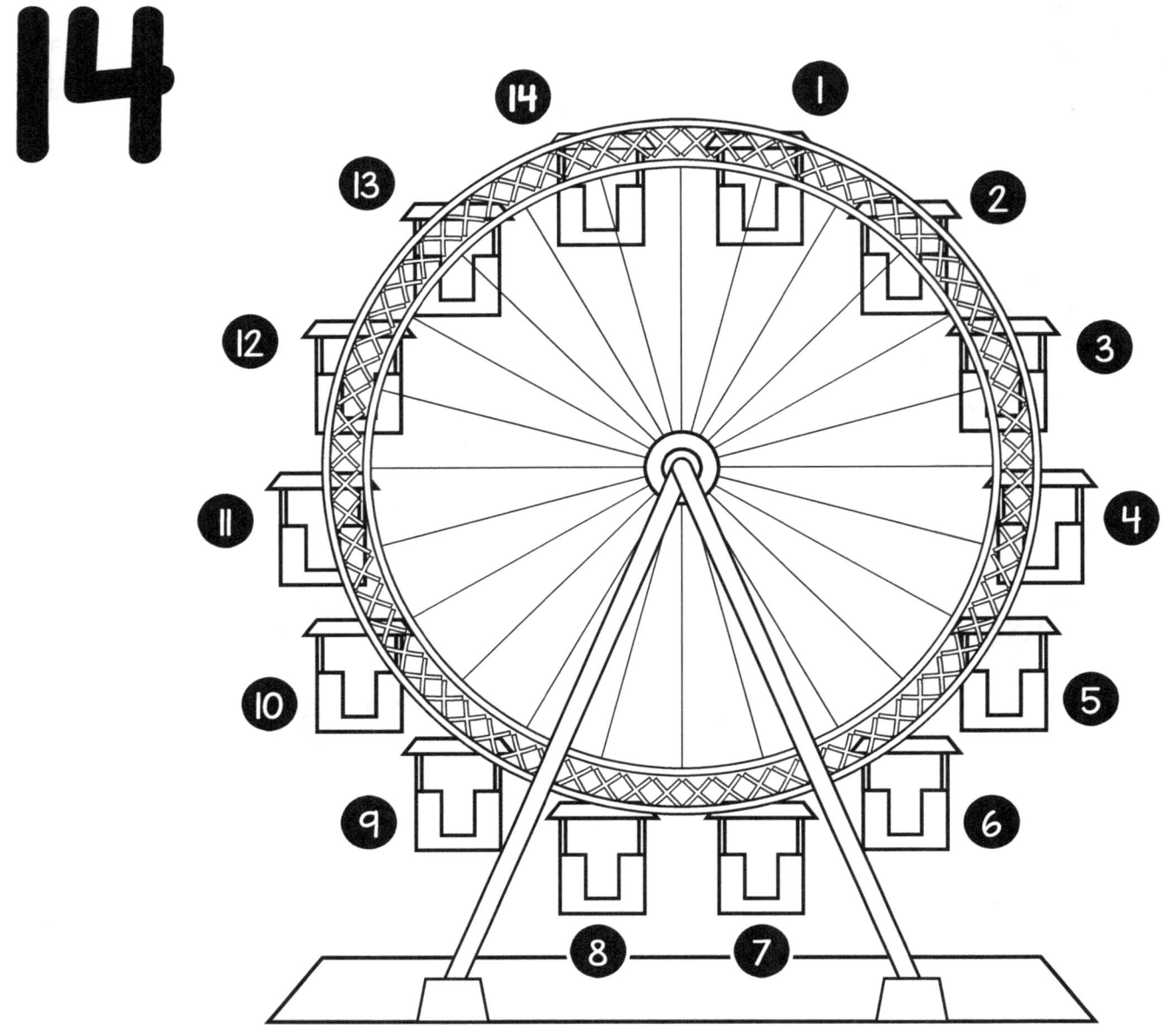

◎ 1 2 3 4 5 6 7 8 9 10 11 12 13 14 15 16 17 18 19 20

THE FERRIS WHEEL HAS 14 CHAIRS.

FERRIS WHEELS HAVE LARGE ELECTRIC MOTORS THAT SPIN THE WHEEL AROUND.

15

0 1 2 3 4 5 6 7 8 9 10 11 12 13 14 15 16 17 18 19 20

THE DUMP TRUCK UNLOADED 15 ROCKS.

SOME DUMP TRUCKS ARE SO LARGE THAT STAIRS ARE ATTACHED TO THE FRONT SO YOU CAN CLIMB UP INTO THE TRUCK.

The Little Engineer Coloring Book: Numbers

16

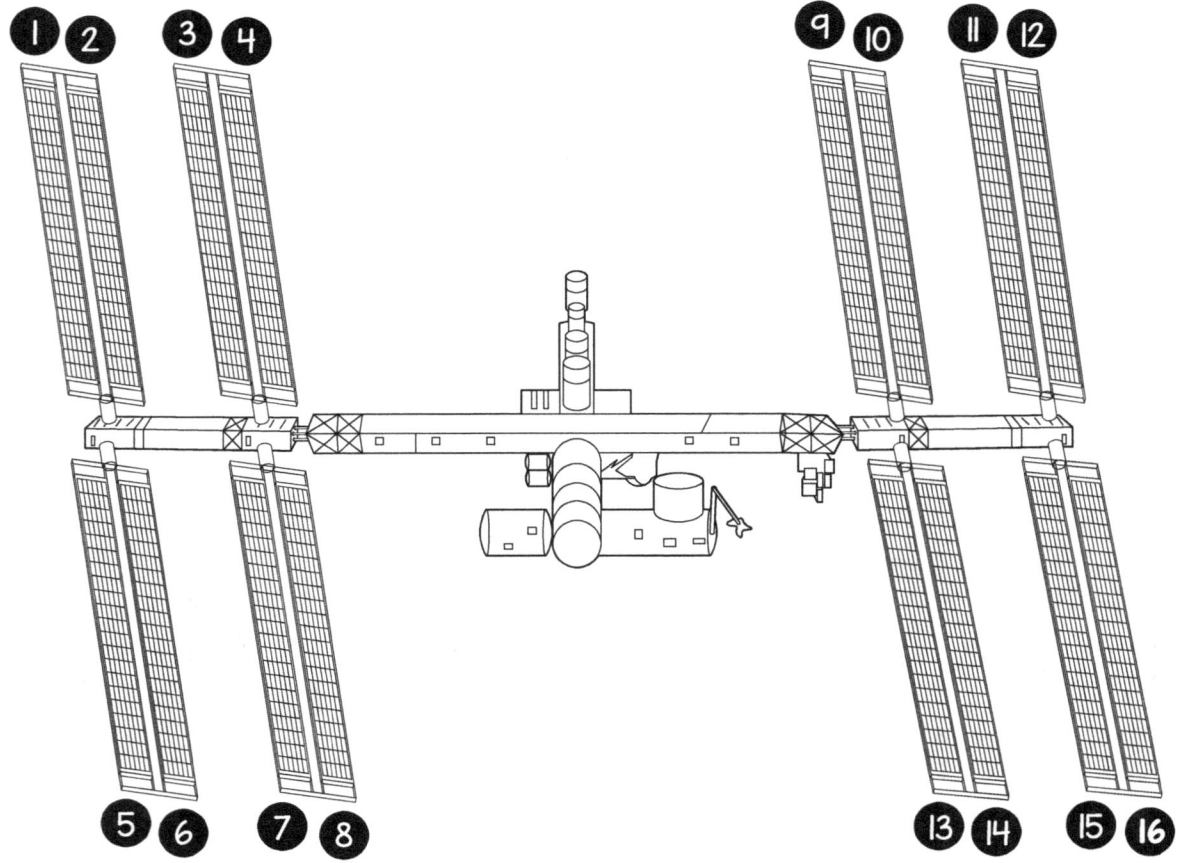

◎ 1 2 3 4 5 6 7 8 9 10 11 12 13 14 15 16 17 18 19 20

THE SPACE STATION HAS 16 SOLAR PANELS.

SOLAR PANELS CAN TURN ENERGY FROM THE SUN INTO ELECTRICITY WHEN THE SUN IS SHINING.

17

◎ 1 2 3 4 5 6 7 8 9 1◎ 11 12 13 14 15 16 17 18 19 2◎

THE CARGO SHIP IS CARRYING 17 CONTAINERS.

GIANT CARGO SHIPS TRANSPORT GOODS
ALL AROUND THE WORLD.

18

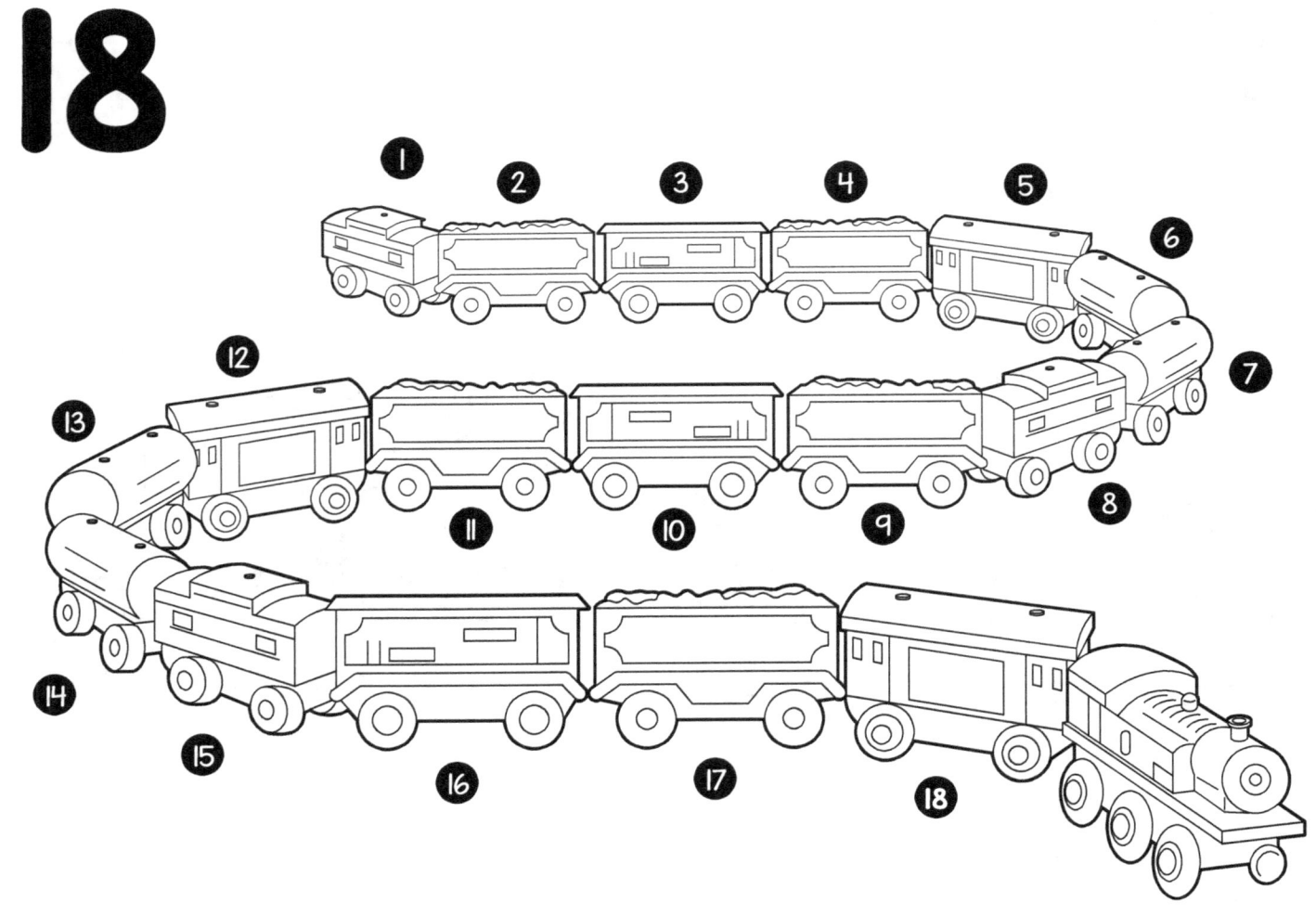

0 1 2 3 4 5 6 7 8 9 10 11 12 13 14 15 16 17 18 19 20

THE TRAIN IS PULLING 18 RAILROAD CARS.

SOME TRAINS CARRY
OVER 50 RAILROAD CARS!

19

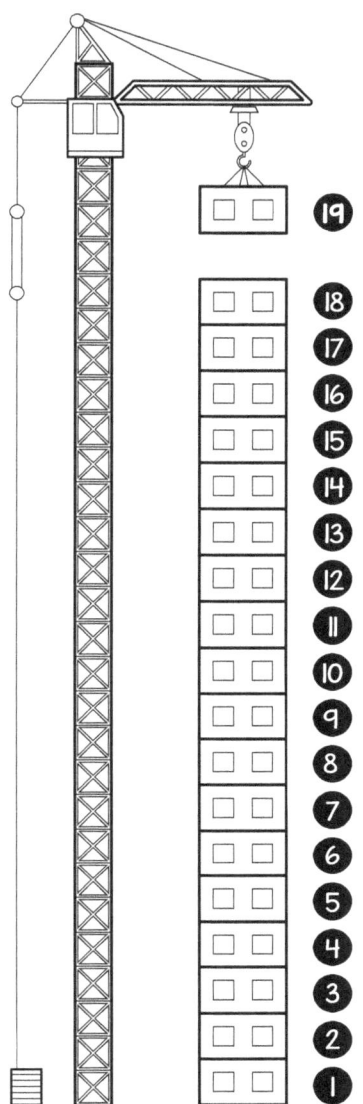

0 1 2 3 4 5 6 7 8 9 10 11 12 13 14 15 16 17 18 19 20

THE BUILDING WILL BE 19 STORIES HIGH.

THE CRANE IS WORKING TO BUILD THE 19TH STORY. SOME BUILDINGS HAVE OVER 100 STORIES!

20

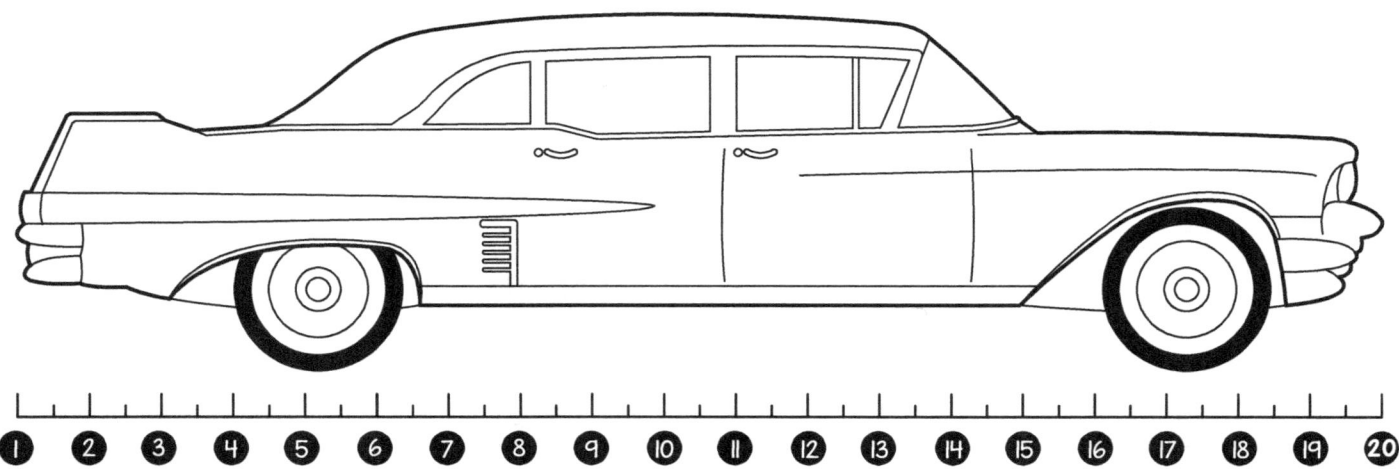

0 1 2 3 4 5 6 7 8 9 10 11 12 13 14 15 16 17 18 19 20

CARS CAN BE 20 FEET LONG.

MOST CARS ARE 15 FEET LONG, BUT SOME STRETCH ALL THE WAY TO 20 FEET.

20 FEET = 6.1 METERS

60 MPH
(miles per hour)

◎ 1 2 3 4 5 6 7 8 9 10 11 12 13 14 15 16 17 18 19 20

CARS DRIVE ABOUT 60 MPH ON THE HIGHWAY.

SUPER CARS CAN EASILY GO FASTER THAN 200 MPH.

60 MPH = 96 KM/H (KILOMETERS PER HOUR)

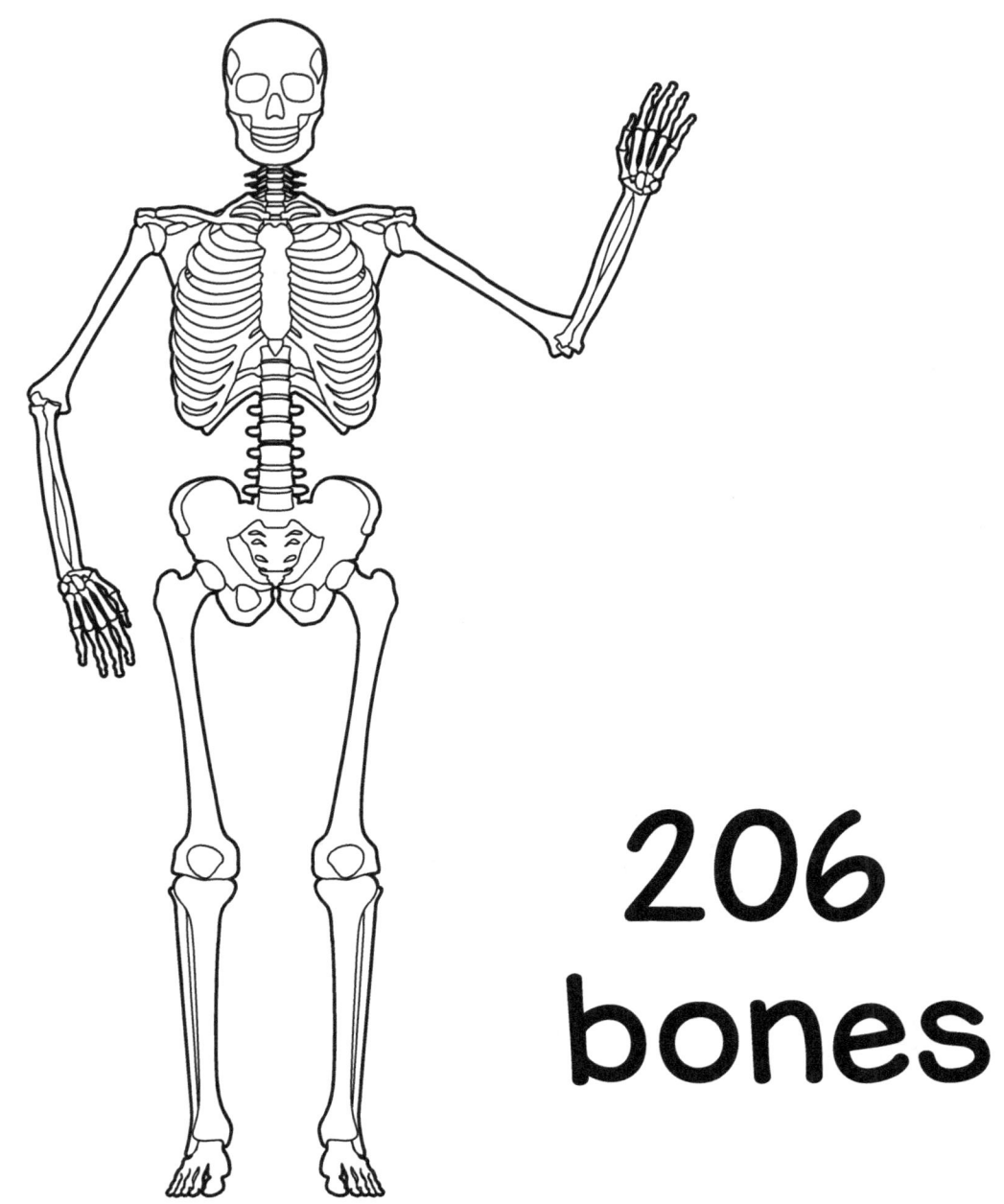

206 bones

0 1 2 3 4 5 6 7 8 9 10 11 12 13 14 15 16 17 18 19 20

THERE ARE 206 BONES IN THE HUMAN BODY.

EACH HAND HAS 27 BONES,
WHILE A FOOT HAS 26 BONES.

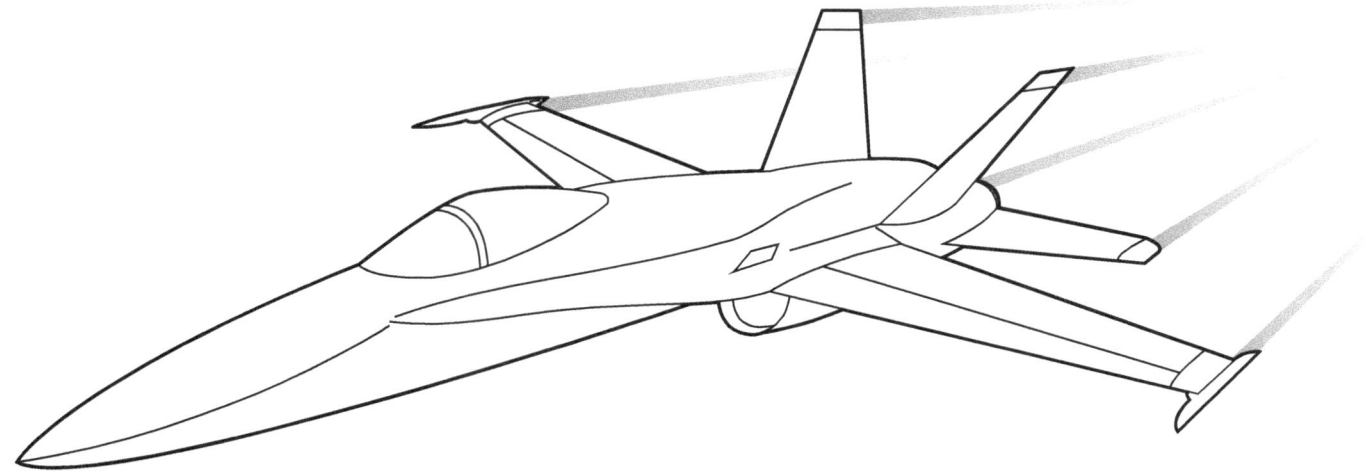

767 MPH

0 1 2 3 4 5 6 7 8 9 10 11 12 13 14 15 16 17 18 19 20

SOUND TRAVELS AT 767 MPH.

IF YOU WERE 767 MILES AWAY FROM A LOUD NOISE, IT WOULD TAKE ONE HOUR FOR THE SOUND TO REACH YOU. JETS CAN FLY SEVERAL TIMES FASTER THAN THIS.

767 MPH = 1,234 KM/H (KILOMETERS PER HOUR)

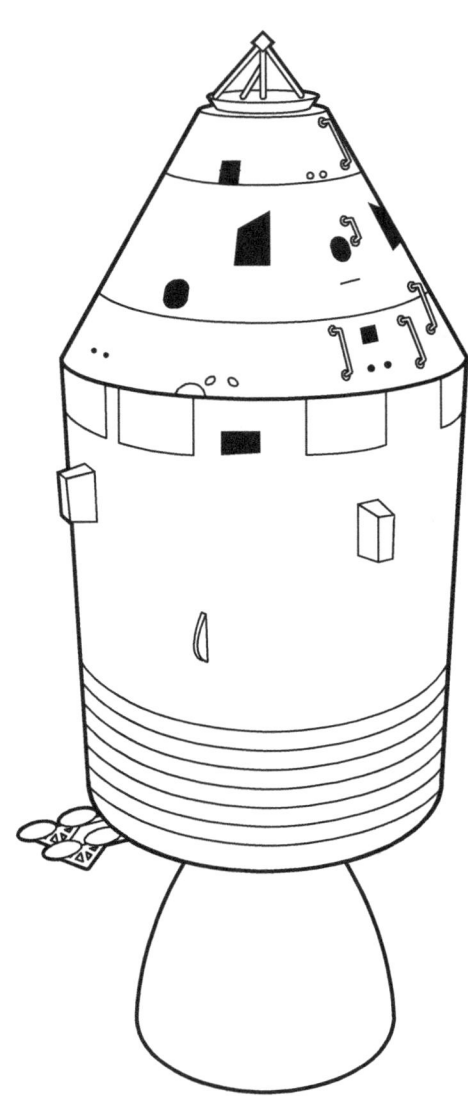

24,790 MPH

◎ 1 2 3 4 5 6 7 8 9 10 11 12 13 14 15 16 17 18 19 20

THE FASTEST HUMANS HAVE TRAVELED 24,790 MPH.

ASTRONAUTS FLYING BACK FROM THE MOON WENT THIS SPEED AS THEY ZOOMED TOWARDS EARTH.

24,790 MPH = 38,896 KM/H (KILOMETERS PER HOUR)

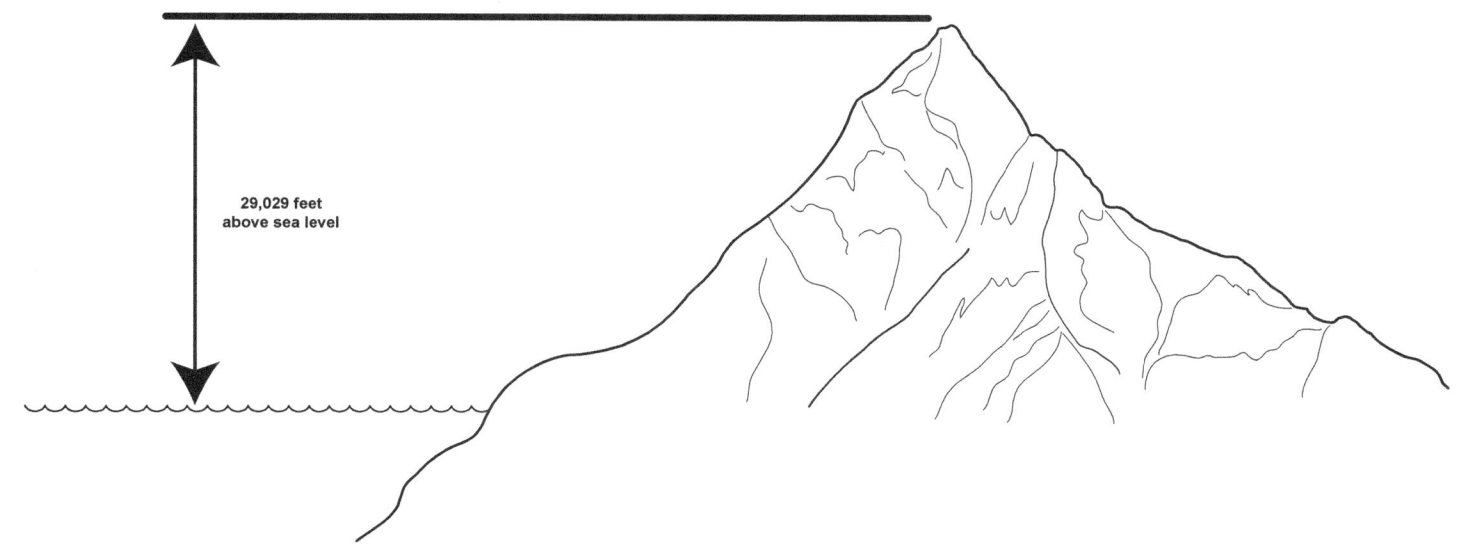

29,029 feet

◎ 1 2 3 4 5 6 7 8 9 1◎ 11 12 13 14 15 16 17 18 19 2◎

MOUNT EVEREST IS 29,029 FEET TALL.

MOUNT EVEREST IS NOT NEXT TO THE OCEAN, BUT MOUNTAINS ARE MEASURED FROM THE OCEAN SURFACE TO THE TOP OF THE MOUNTAIN.

29,029 FEET = 8,848 METERS

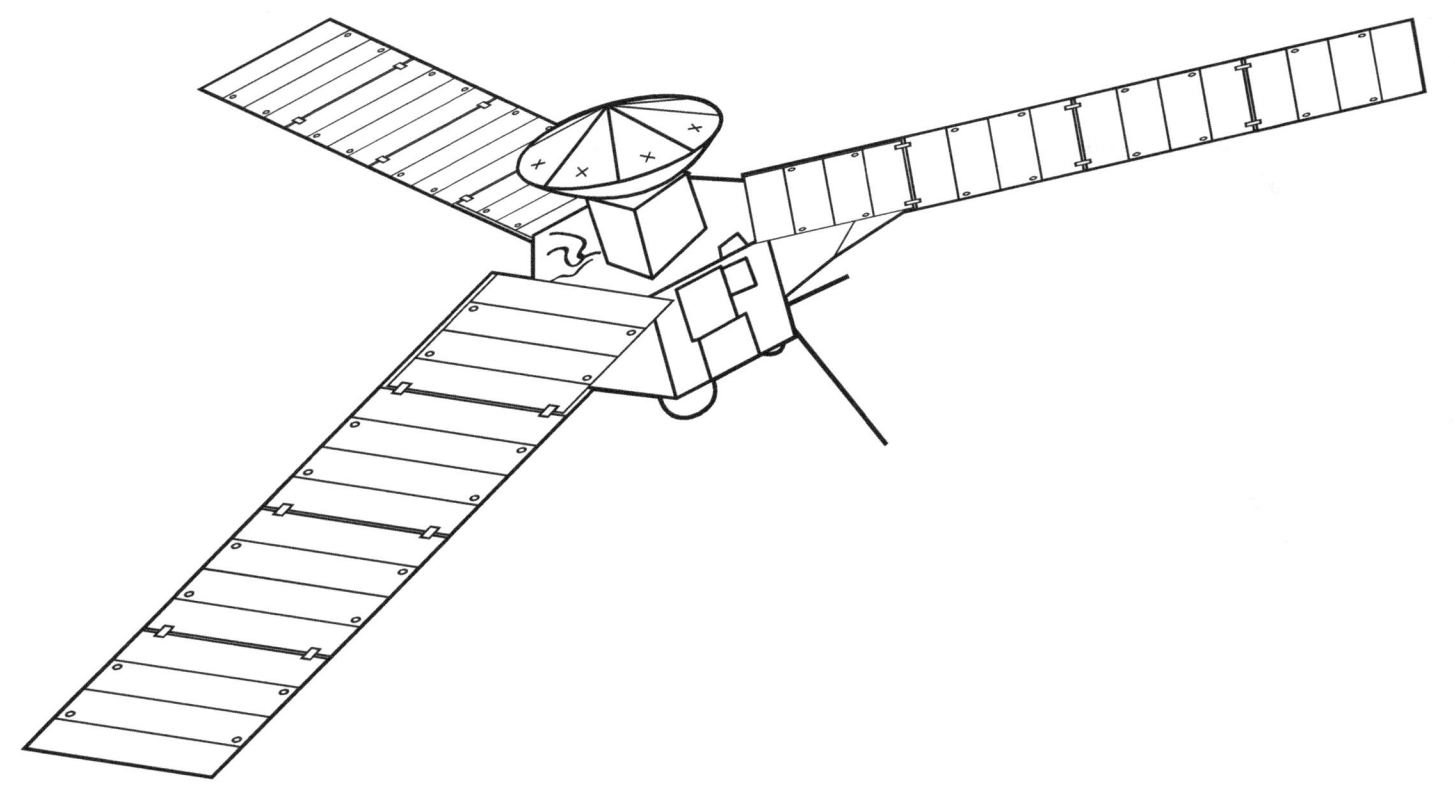

160,000 MPH

0 1 2 3 4 5 6 7 8 9 10 11 12 13 14 15 16 17 18 19 20

THE FASTEST SPACE PROBE WENT 160,000 MPH.

SPACE PROBES ARE SCIENCE EXPERIMENTS WE HAVE SENT OUT INTO SPACE TO STUDY OTHER PLANETS OR THE SUN.

160,000 MPH = 257,495 KM/H (KILOMETERS PER HOUR)

670,616,629 MPH

◎ 1 2 3 4 5 6 7 8 9 10 11 12 13 14 15 16 17 18 19 20

LIGHT TRAVELS AT 670,616,629 MPH.

WOW, THAT'S FAST! LIGHT IS INCREDIBLY FAST.
WE ARE UNSURE IF ANYTHING TRAVELS
FASTER THAN THE SPEED OF LIGHT.

670,616,629 MPH = 1,079,252,848 KM/H (KILOMETERS PER HOUR)

Double the Fun!

Each image repeated to give your little engineer a chance to color each page again!

1

◎ 1 2 3 4 5 6 7 8 9 10 11 12 13 14 15 16 17 18 19 20

THERE IS 1 EARTH.

EARTH IS THE PLANET WE LIVE ON, AND EARTH IS THE BEST PLACE TO LIVE IN OUR SOLAR SYSTEM.

2

0 1 2 3 4 5 6 7 8 9 10 11 12 13 14 15 16 17 18 19 20

A MOTORCYCLE HAS 2 WHEELS.

BICYCLES ALSO HAVE 2 WHEELS, BUT TRICYCLES HAVE 3 WHEELS.

3

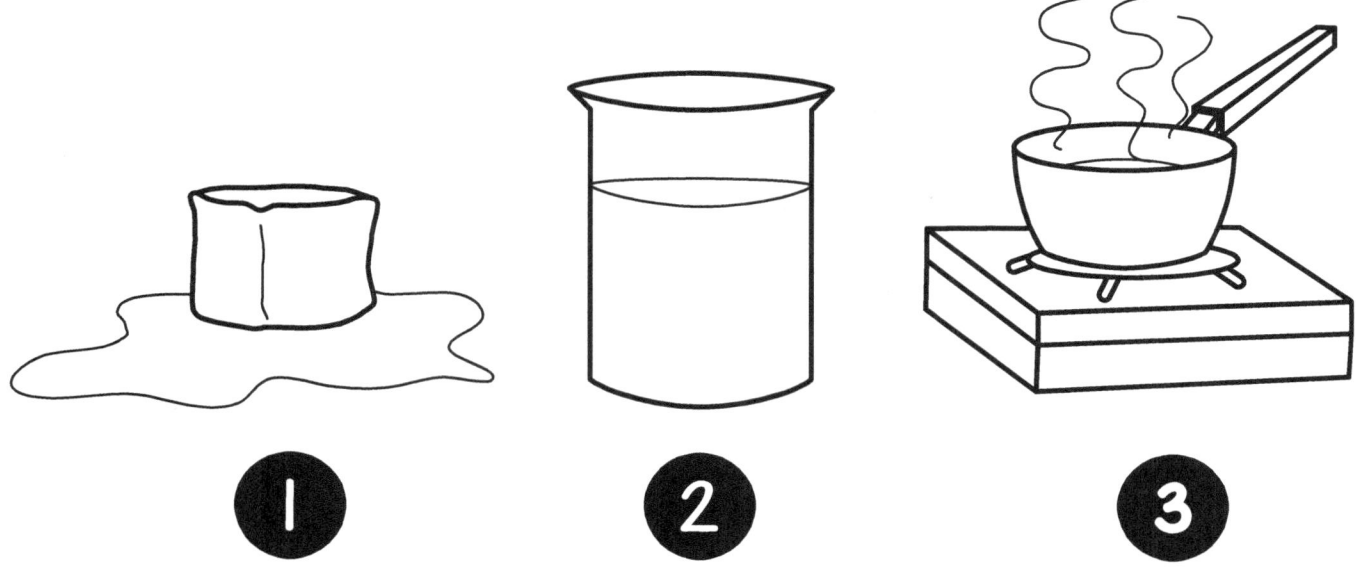

0 1 2 3 4 5 6 7 8 9 10 11 12 13 14 15 16 17 18 19 20

WATER CAN COME IN 3 FORMS.

WATER CAN BE SOLID (LIKE ICE CUBES), LIQUID (LIKE THE WATER WE DRINK) OR GAS (LIKE STEAM WHEN COOKING SPAGHETTI).

4

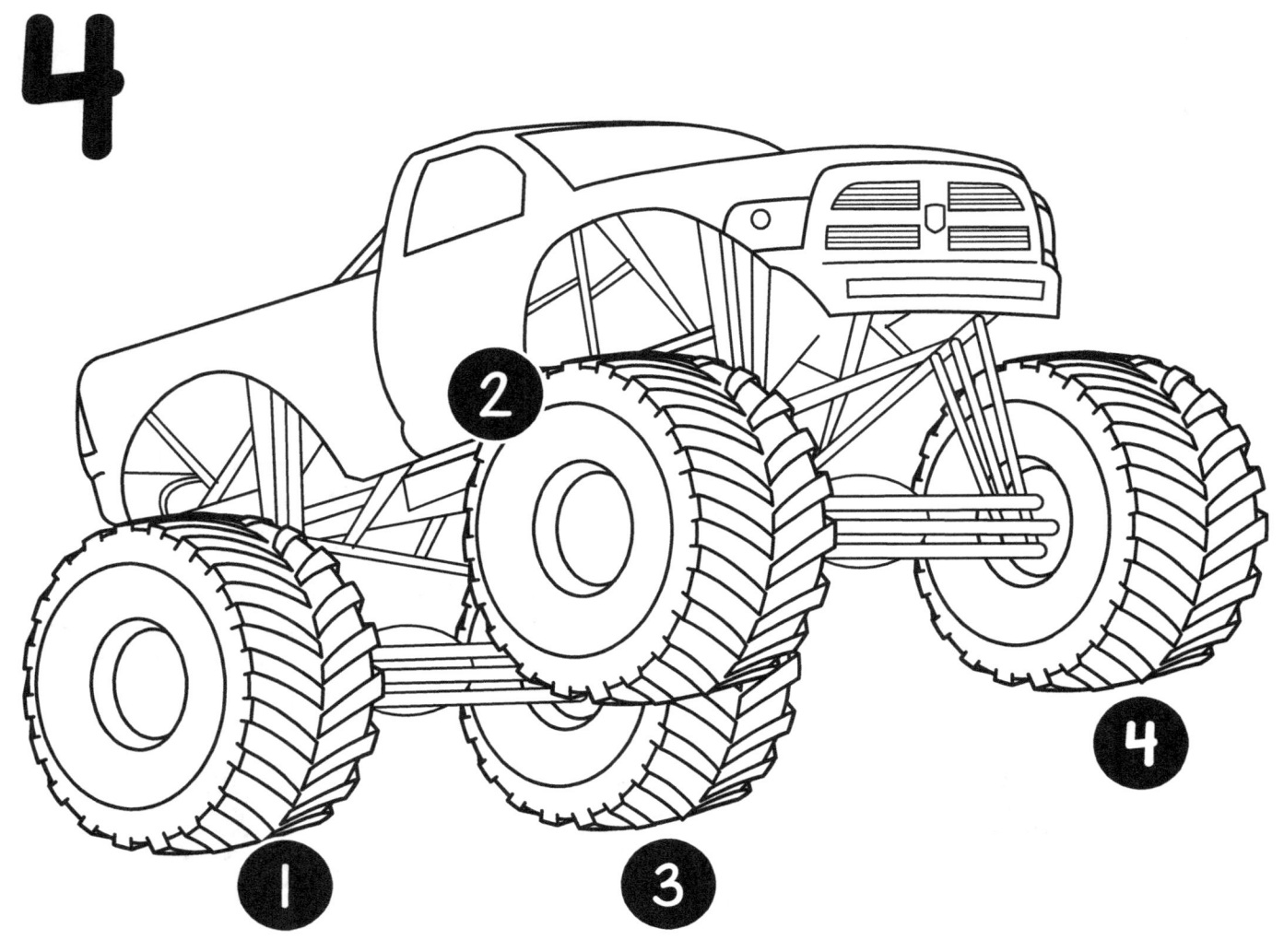

◎ 1 2 3 4 5 6 7 8 9 10 11 12 13 14 15 16 17 18 19 20

A MONSTER TRUCK HAS 4 HUGE TIRES.

THESE LARGE TIRES HELP THE TRUCK RUN OVER CARS!

5

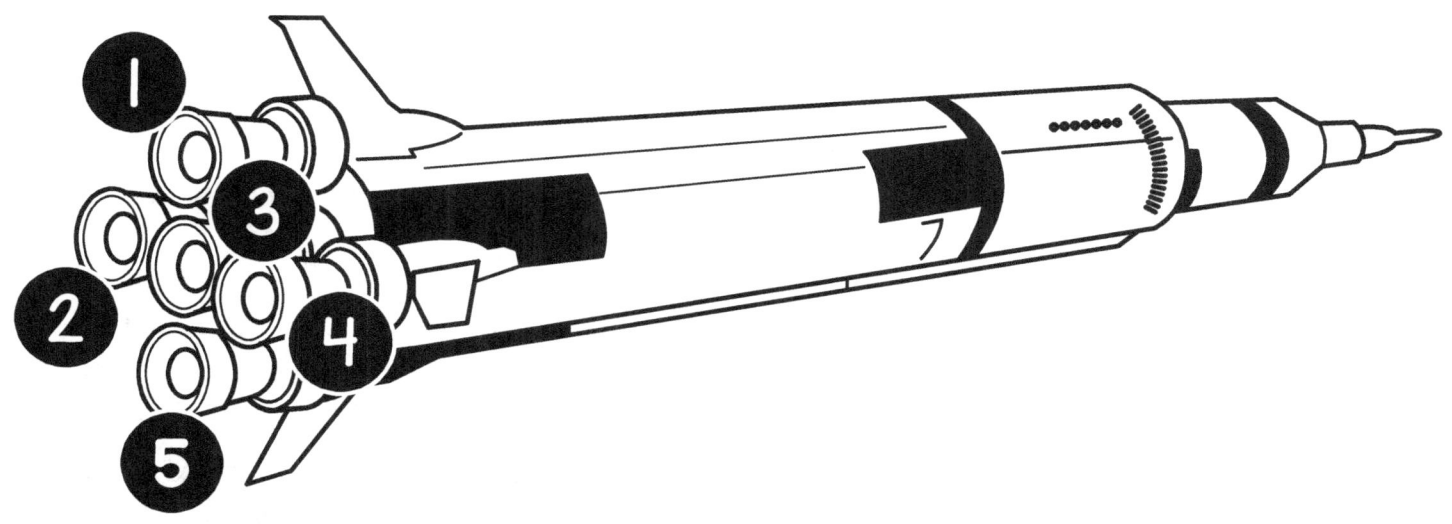

◎ 1 2 3 4 5 6 7 8 9 1◎ 11 12 13 14 15 16 17 18 19 2◎

THE SATURN V (5) HAS 5 POWERFUL ROCKET ENGINES.

THE SATURN V TOOK ASTRONAUTS TO THE MOON 50 YEARS AGO!

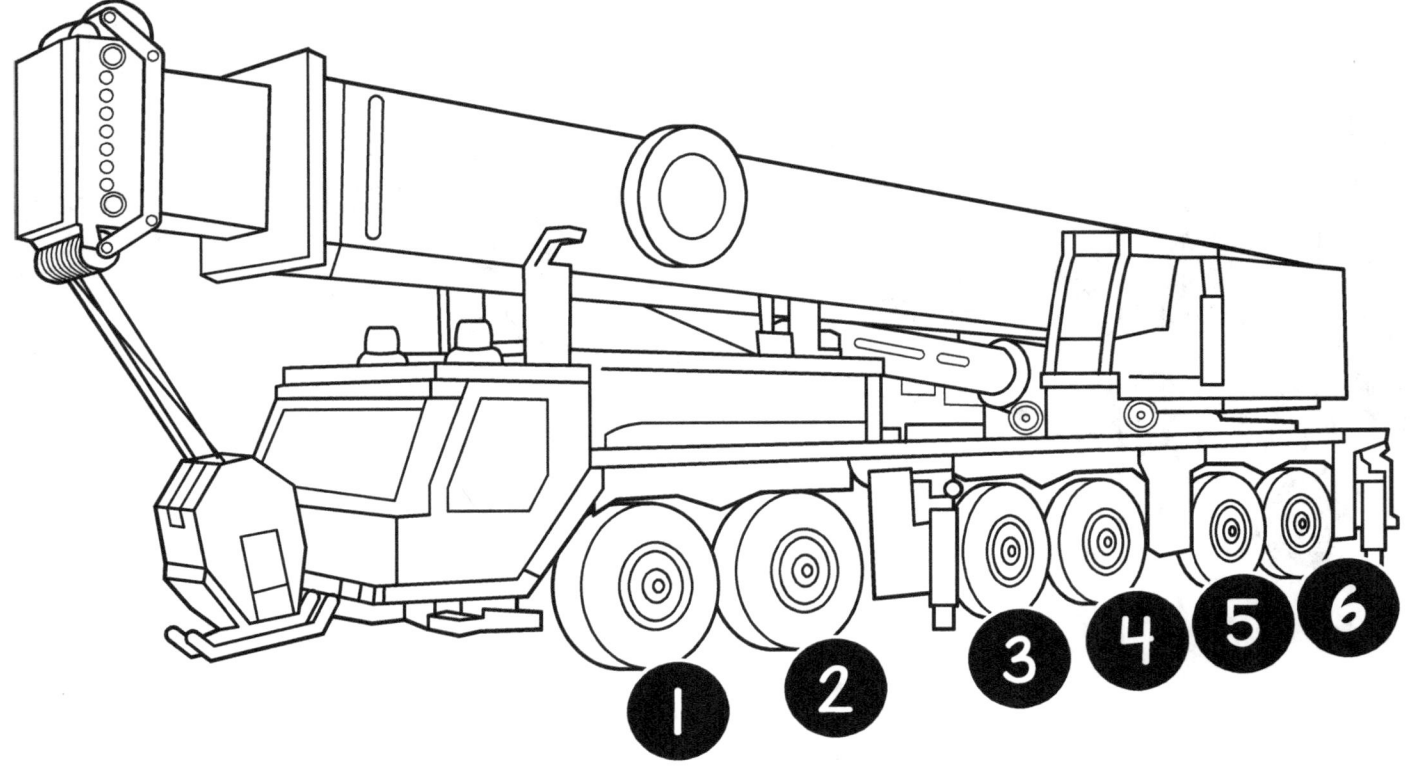

0 1 2 3 4 5 6 7 8 9 10 11 12 13 14 15 16 17 18 19 20

THE CRANE HAS 6 WHEELS ON EACH SIDE.

THIS MEGA CRANE IS INCREDIBLY POWERFUL, AND IS USED TO BUILD LARGE BUILDINGS AND BRIDGES.

7

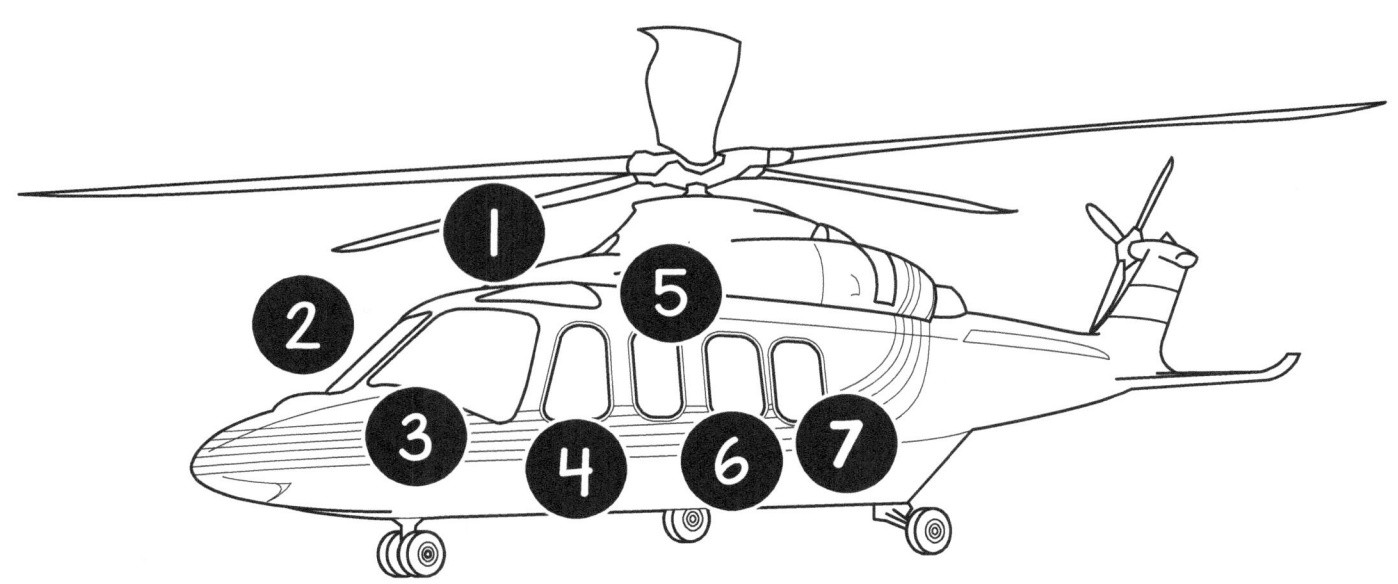

◎ 1 2 3 4 5 6 7 8 9 10 11 12 13 14 15 16 17 18 19 20

THE HELICOPTER HAS 7 WINDOWS.

HELICOPTER BLADES ACT LIKE A BIG FAN
AND BLOW AIR DOWN WHICH
MAKES THE HELICOPTER RAISE UP IN THE SKY.

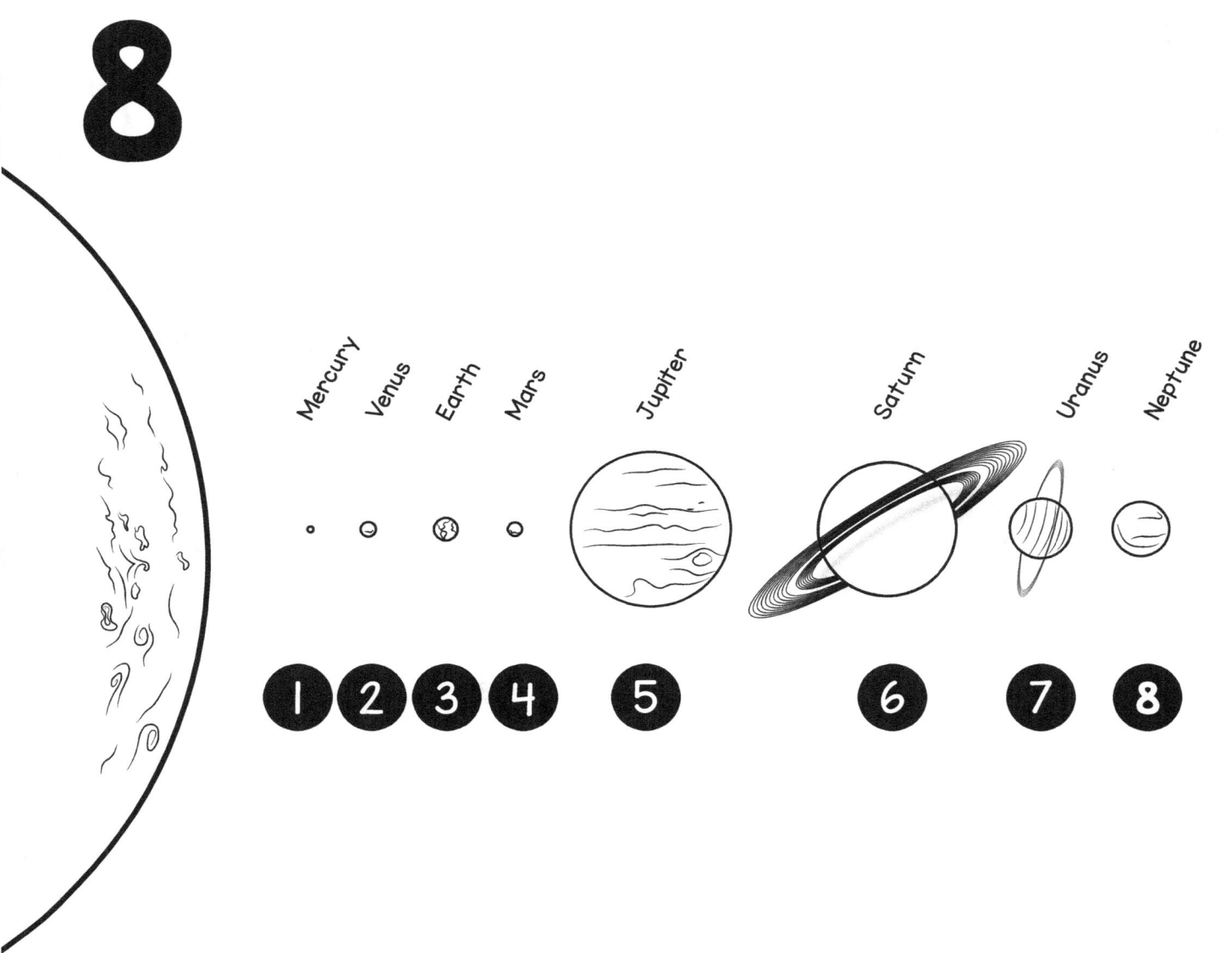

THERE ARE 8 PLANETS IN OUR SOLAR SYSTEM.

ALL OF THE PLANETS ORBIT (OR GO AROUND) THE SUN. THE PLANETS ARE FARTHER APART THAN SHOWN HERE.

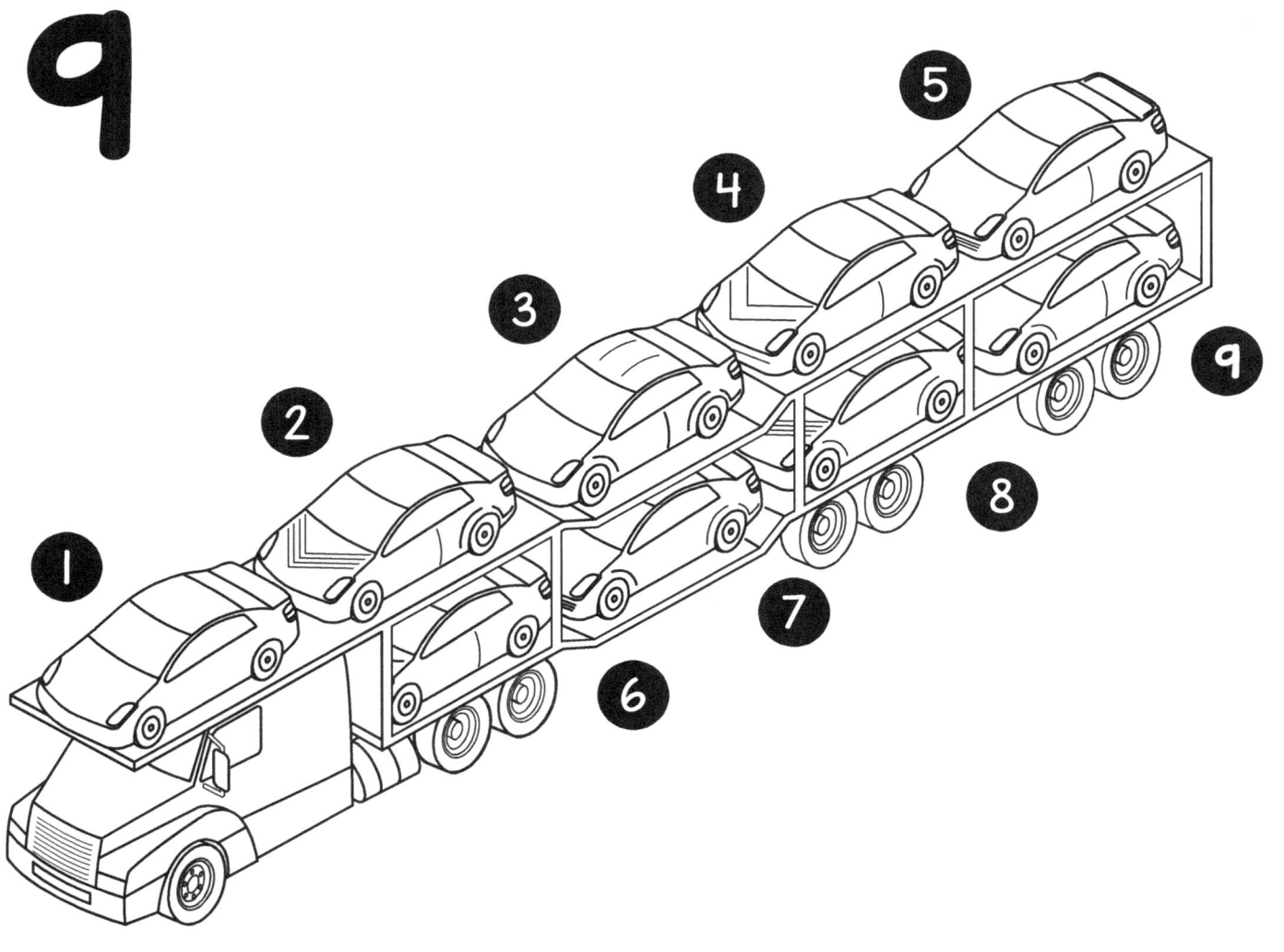

0 1 2 3 4 5 6 7 8 9 10 11 12 13 14 15 16 17 18 19 20

THE TRANSPORT TRUCK IS CARRYING 9 CARS.

BIG DIESEL TRUCKS ARE VERY POWERFUL, AND CAN CARRY MANY CARS.

10

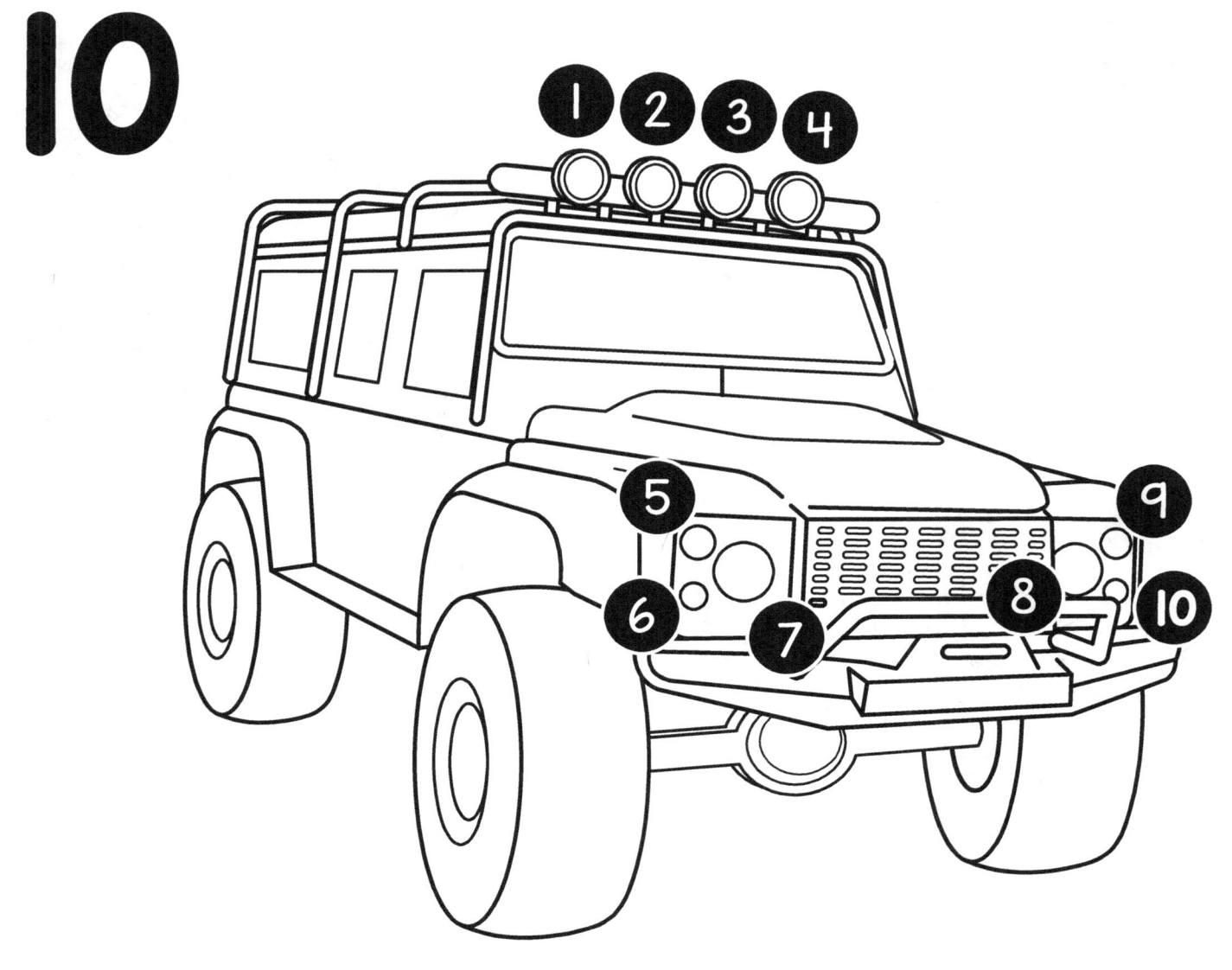

THE OFF-ROAD VEHICLE HAS 10 LIGHTS.

THE LIGHTS HELP MAKE SURE THE DRIVER CAN SEE WHEN DRIVING AT NIGHT.

0 1 2 3 4 5 6 7 8 9 10 11 12 13 14 15 16 17 18 19 20

THE TRACTOR IS PULLING 11 BALES OF HAY.

HAY BALES ARE MADE FROM CUT GRASS FROM THE FIELD, AND THEY ARE USED TO FEED THE COWS.

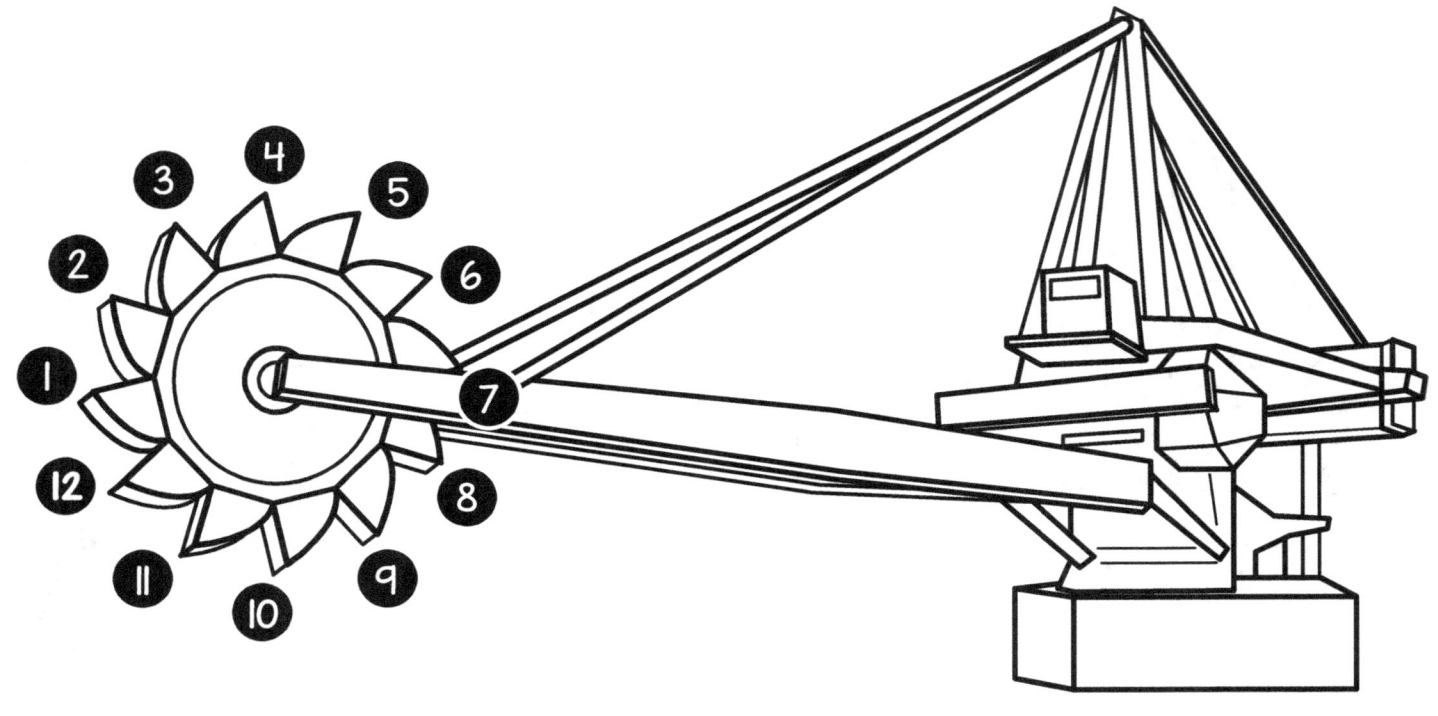

0 1 2 3 4 5 6 7 8 9 10 11 12 13 14 15 16 17 18 19 20

THE BUCKET WHEEL EXCAVATOR HAS 12 BUCKETS.

THIS EXCAVATOR IS BIGGER THAN YOUR HOUSE! A CAR COULD FIT INSIDE EACH BUCKET!

13

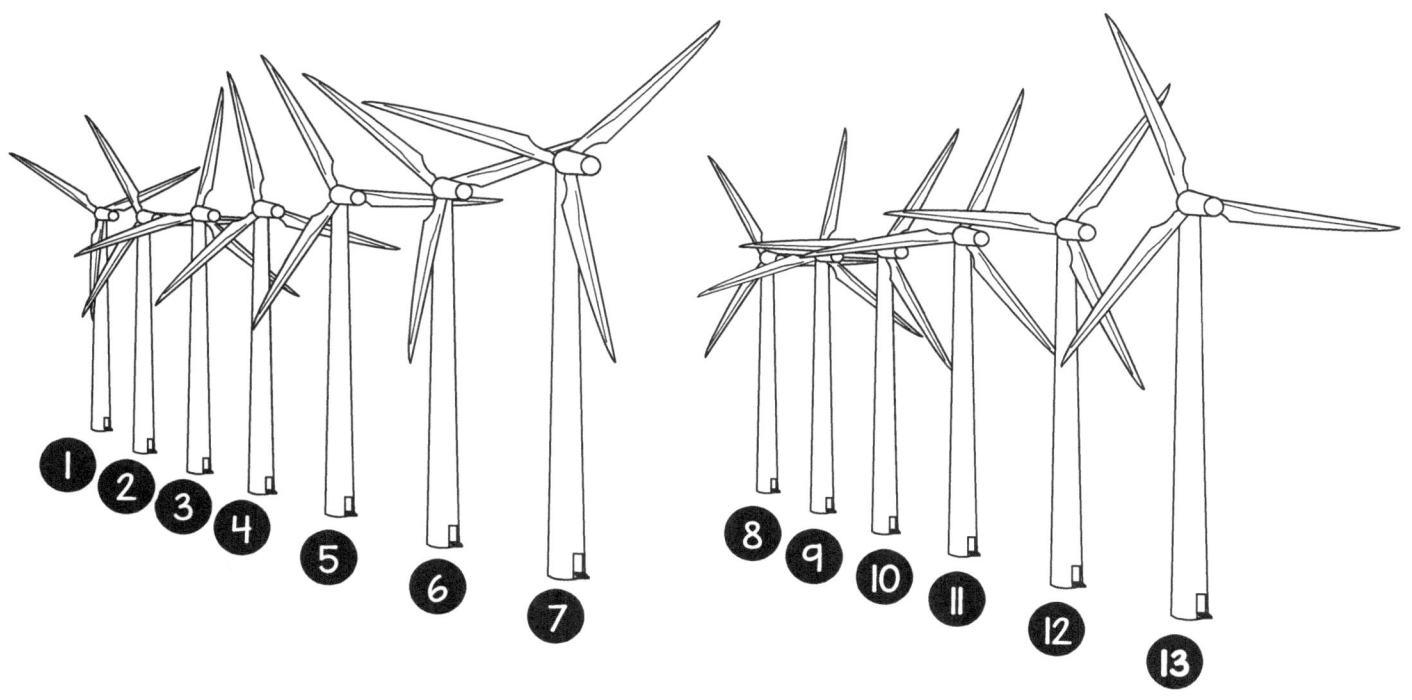

◎ 1 2 3 4 5 6 7 8 9 10 11 12 13 14 15 16 17 18 19 20

THERE ARE 13 WINDMILLS IN THE FIELD.

WINDMILLS GENERATE ELECTRICITY
WHEN STRONG WINDS SPIN THE BLADES.

14

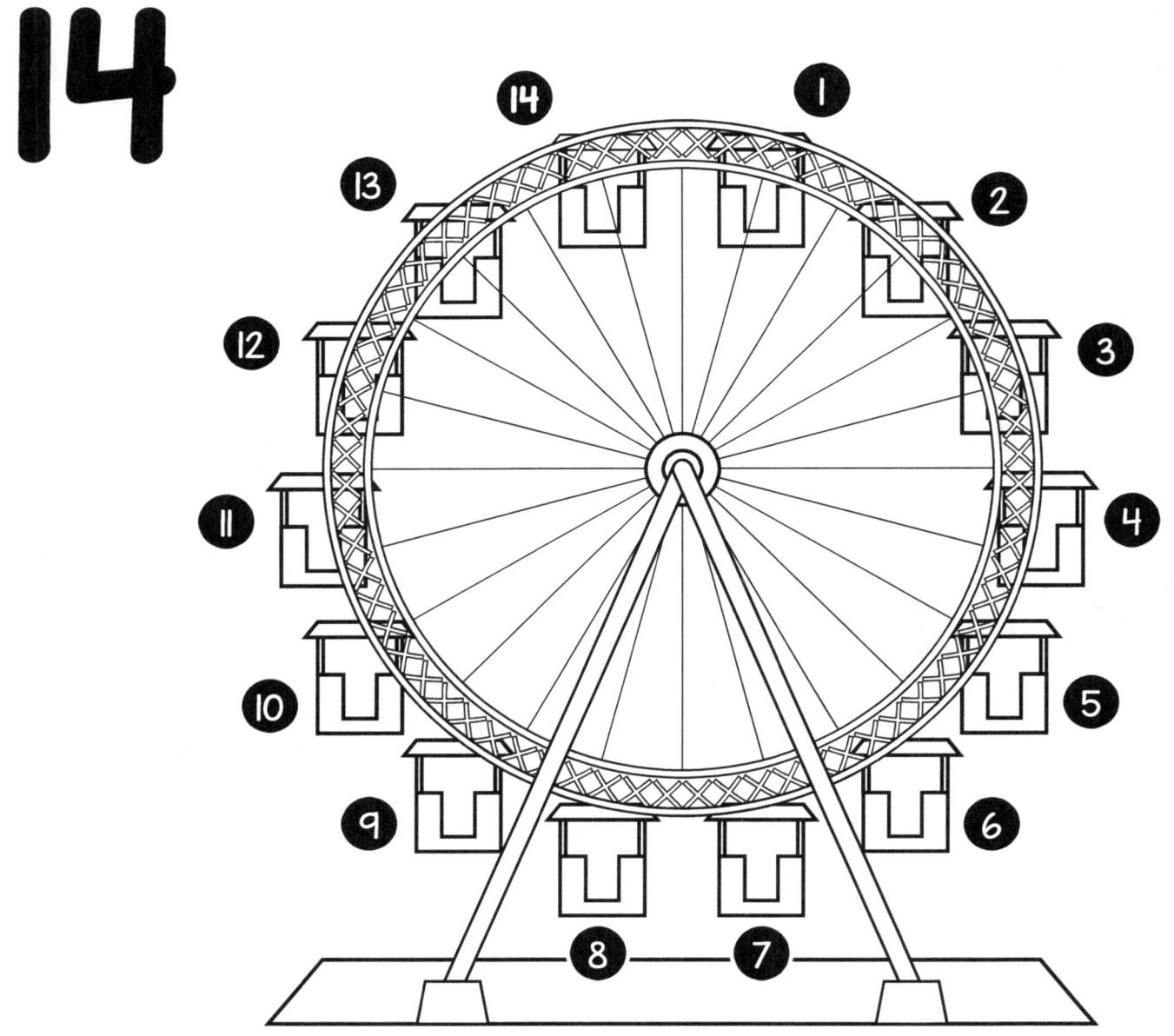

◎ 1 2 3 4 5 6 7 8 9 10 11 12 13 14 15 16 17 18 19 20

THE FERRIS WHEEL HAS 14 CHAIRS.

FERRIS WHEELS HAVE LARGE ELECTRIC MOTORS THAT SPIN THE WHEEL AROUND.

◎ 1 2 3 4 5 6 7 8 9 10 11 12 13 14 15 16 17 18 19 20

THE DUMP TRUCK UNLOADED 15 ROCKS.

SOME DUMP TRUCKS ARE SO LARGE THAT STAIRS ARE ATTACHED TO THE FRONT SO YOU CAN CLIMB UP INTO THE TRUCK.

16

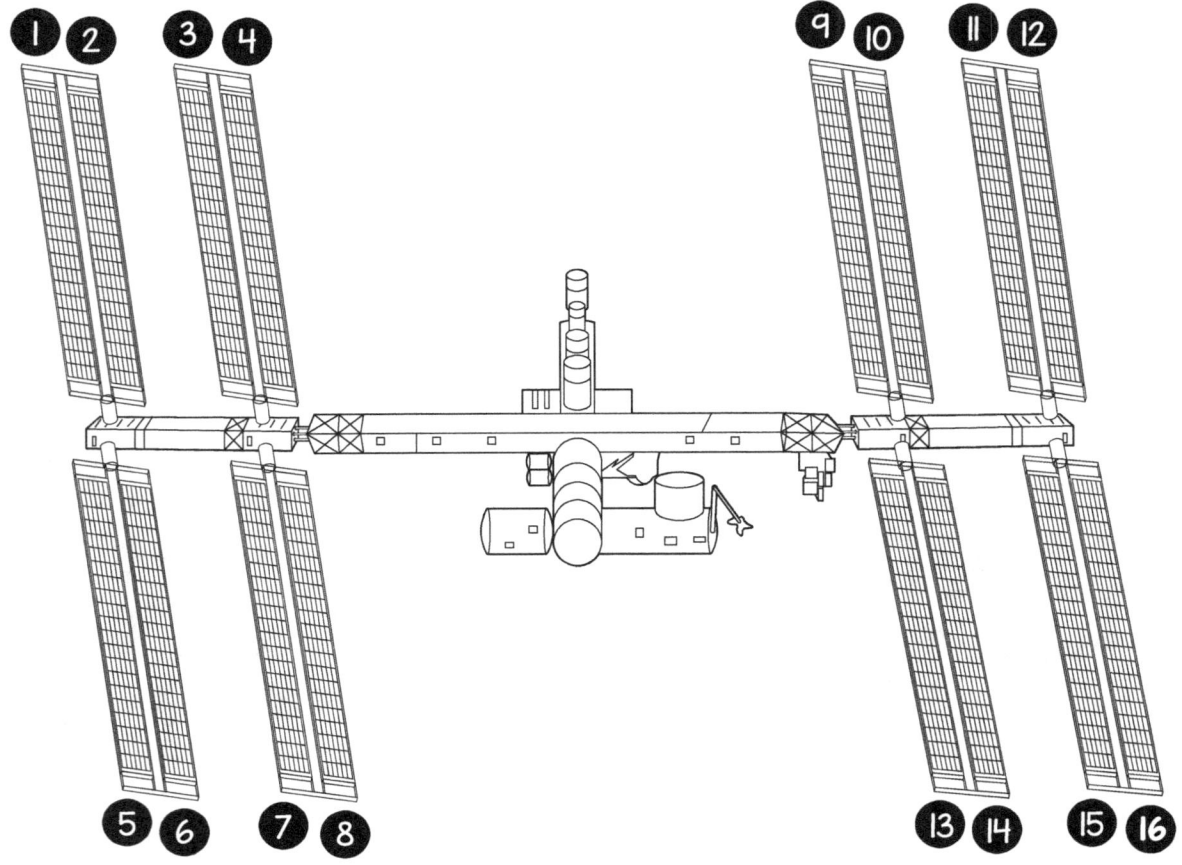

◎ 1 2 3 4 5 6 7 8 9 10 11 12 13 14 15 16 17 18 19 20

THE SPACE STATION HAS 16 SOLAR PANELS.

SOLAR PANELS CAN TURN ENERGY FROM THE SUN INTO ELECTRICITY WHEN THE SUN IS SHINING.

17

0 1 2 3 4 5 6 7 8 9 10 11 12 13 14 15 16 17 18 19 20

THE CARGO SHIP IS CARRYING 17 CONTAINERS.

GIANT CARGO SHIPS TRANSPORT GOODS ALL AROUND THE WORLD.

18

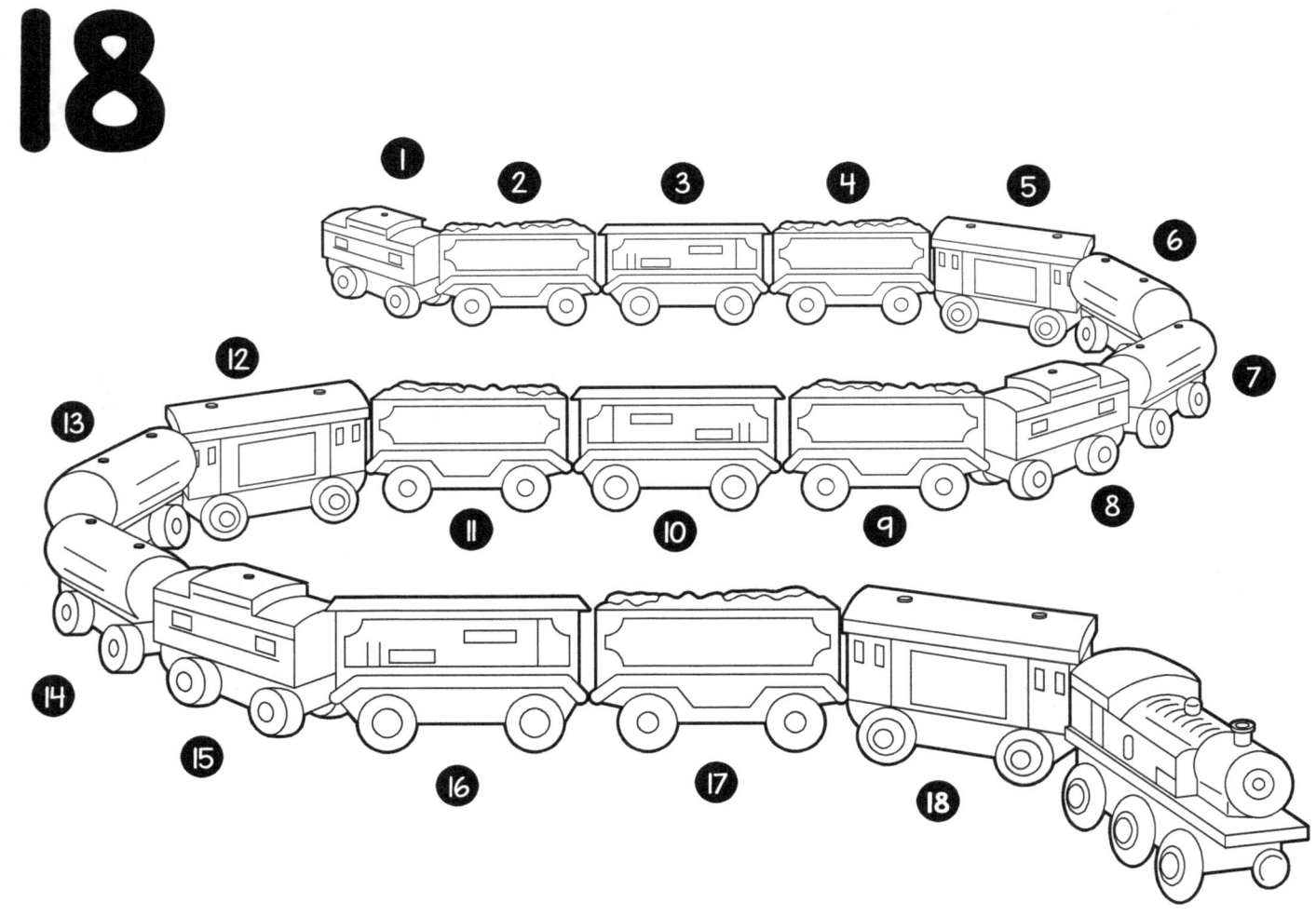

◎ 1 2 3 4 5 6 7 8 9 10 11 12 13 14 15 16 17 18 19 20

THE TRAIN IS PULLING 18 RAILROAD CARS.

SOME TRAINS CARRY OVER 50 RAILROAD CARS!

19

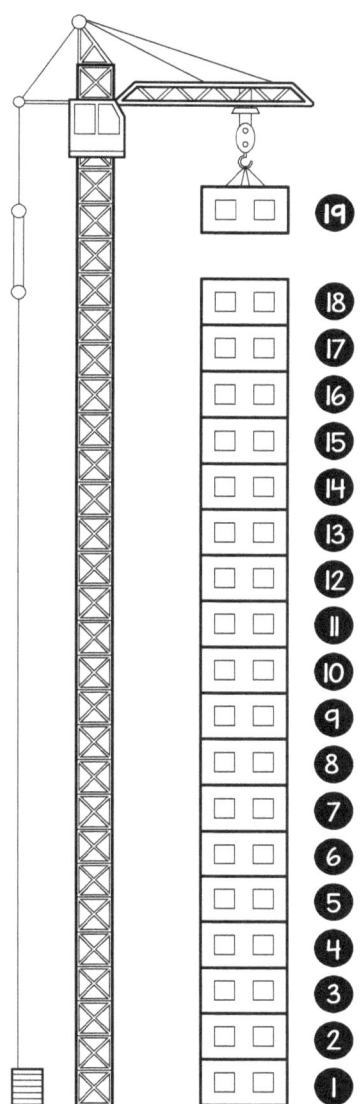

◎ 1 2 3 4 5 6 7 8 9 10 11 12 13 14 15 16 17 18 19 20

THE BUILDING WILL BE 19 STORIES HIGH.

THE CRANE IS WORKING TO BUILD THE 19TH STORY. SOME BUILDINGS HAVE OVER 100 STORIES!

20

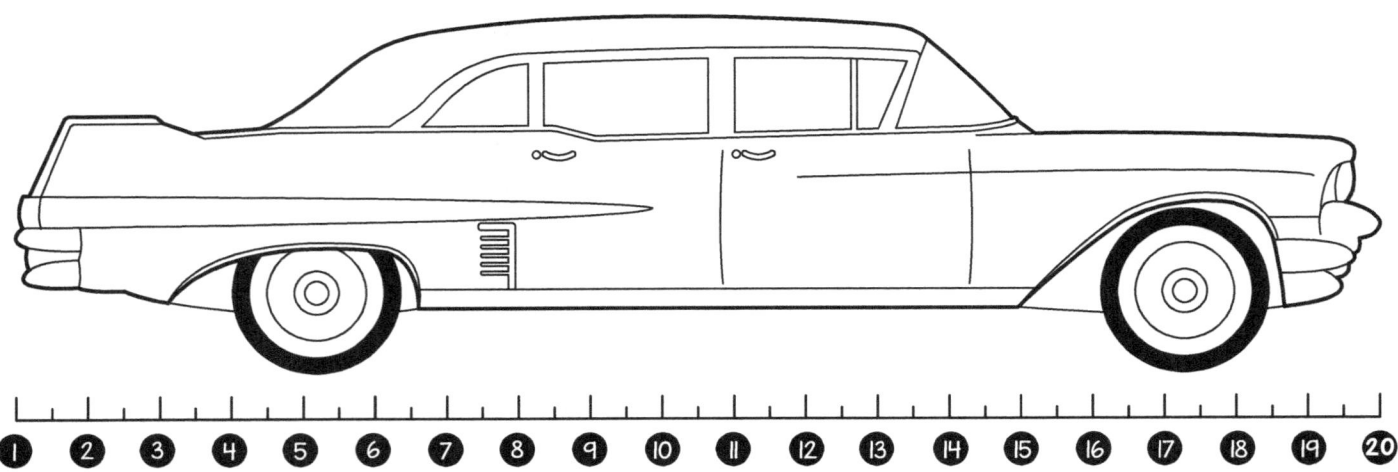

0 1 2 3 4 5 6 7 8 9 10 11 12 13 14 15 16 17 18 19 20

CARS CAN BE 20 FEET LONG.

MOST CARS ARE 15 FEET LONG, BUT SOME STRETCH ALL THE WAY TO 20 FEET.

20 FEET = 6.1 METERS

60 MPH
(miles per hour)

◎ 1 2 3 4 5 6 7 8 9 10 11 12 13 14 15 16 17 18 19 20

CARS DRIVE ABOUT 60 MPH ON THE HIGHWAY.

SUPER CARS CAN EASILY GO FASTER THAN 200 MPH.

60 MPH = 96 KM/H (KILOMETERS PER HOUR)

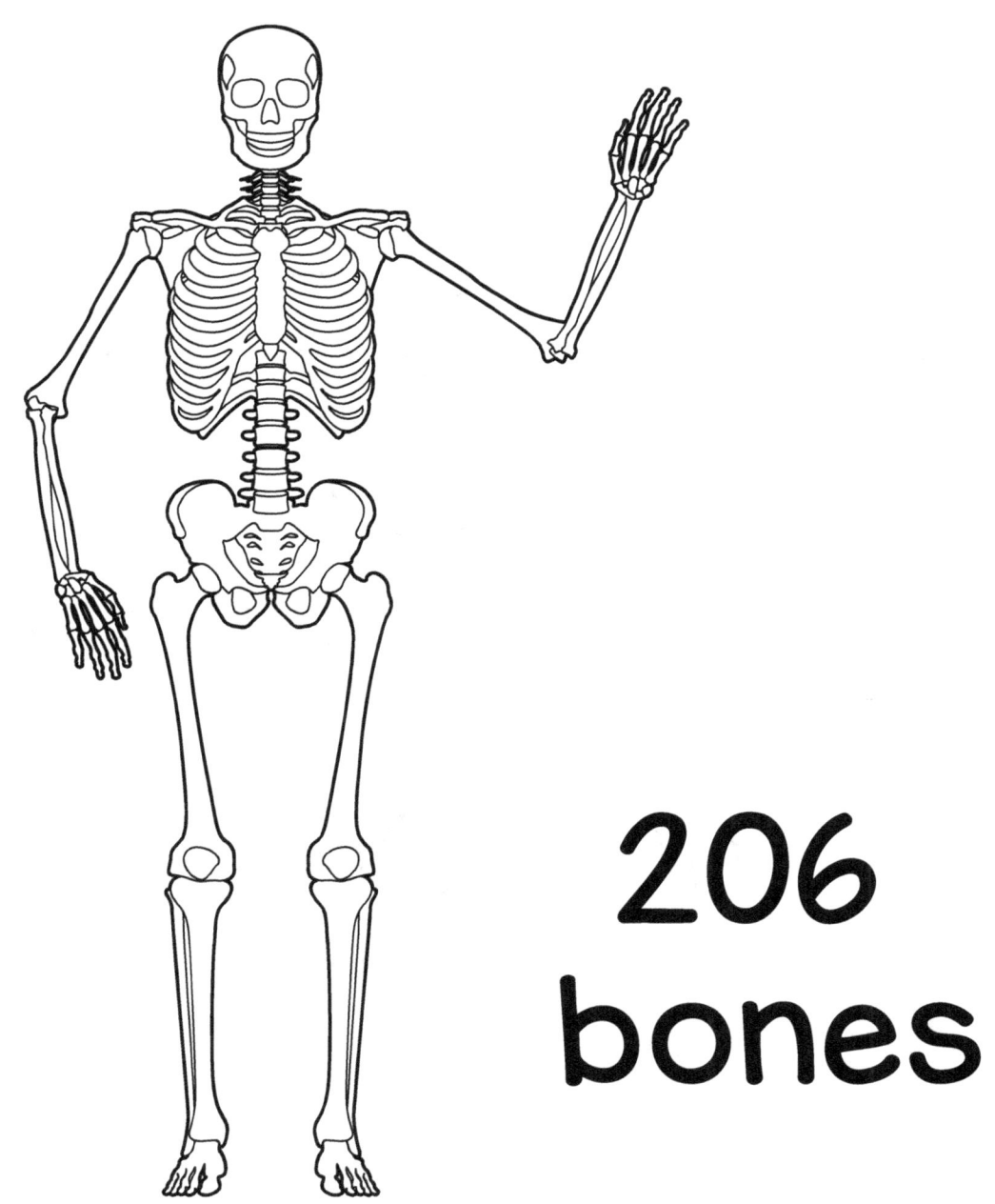

206 bones

⓪ ① ② ③ ④ ⑤ ⑥ ⑦ ⑧ ⑨ ⑩ 11 12 13 14 15 16 17 18 19 2⓪

THERE ARE 206 BONES IN THE HUMAN BODY.

EACH HAND HAS 27 BONES,
WHILE A FOOT HAS 26 BONES.

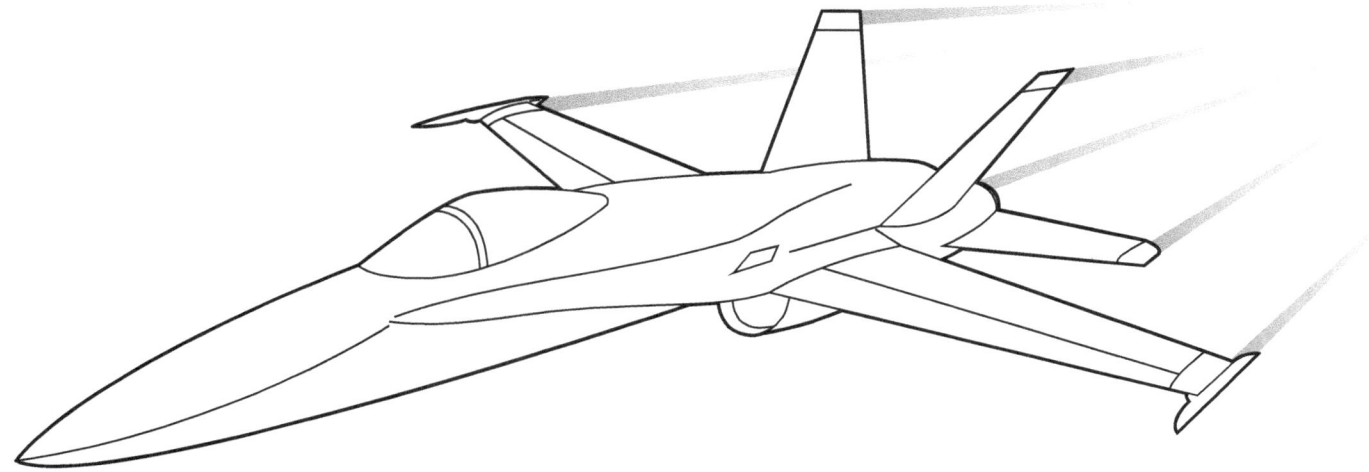

767 MPH

0 1 2 3 4 5 6 7 8 9 10 11 12 13 14 15 16 17 18 19 20

SOUND TRAVELS AT 767 MPH.

IF YOU WERE 767 MILES AWAY FROM A LOUD NOISE, IT WOULD TAKE ONE HOUR FOR THE SOUND TO REACH YOU. JETS CAN FLY SEVERAL TIMES FASTER THAN THIS.

767 MPH = 1,234 KM/H (KILOMETERS PER HOUR)

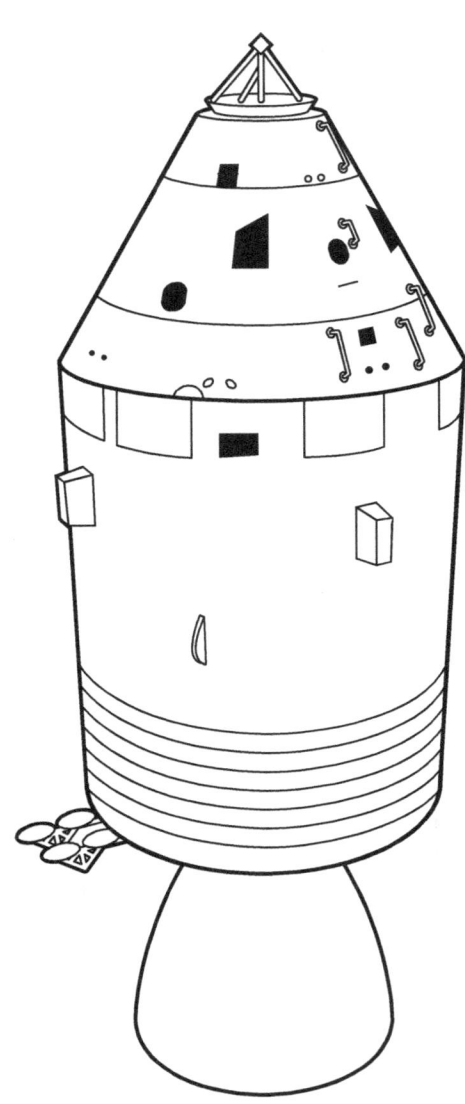

24,790 MPH

0 1 2 3 4 5 6 7 8 9 10 11 12 13 14 15 16 17 18 19 20

THE FASTEST HUMANS HAVE TRAVELED 24,790 MPH.

ASTRONAUTS FLYING BACK FROM THE MOON WENT THIS SPEED AS THEY ZOOMED TOWARDS EARTH.

24,790 MPH = 38,896 KM/H (KILOMETERS PER HOUR)

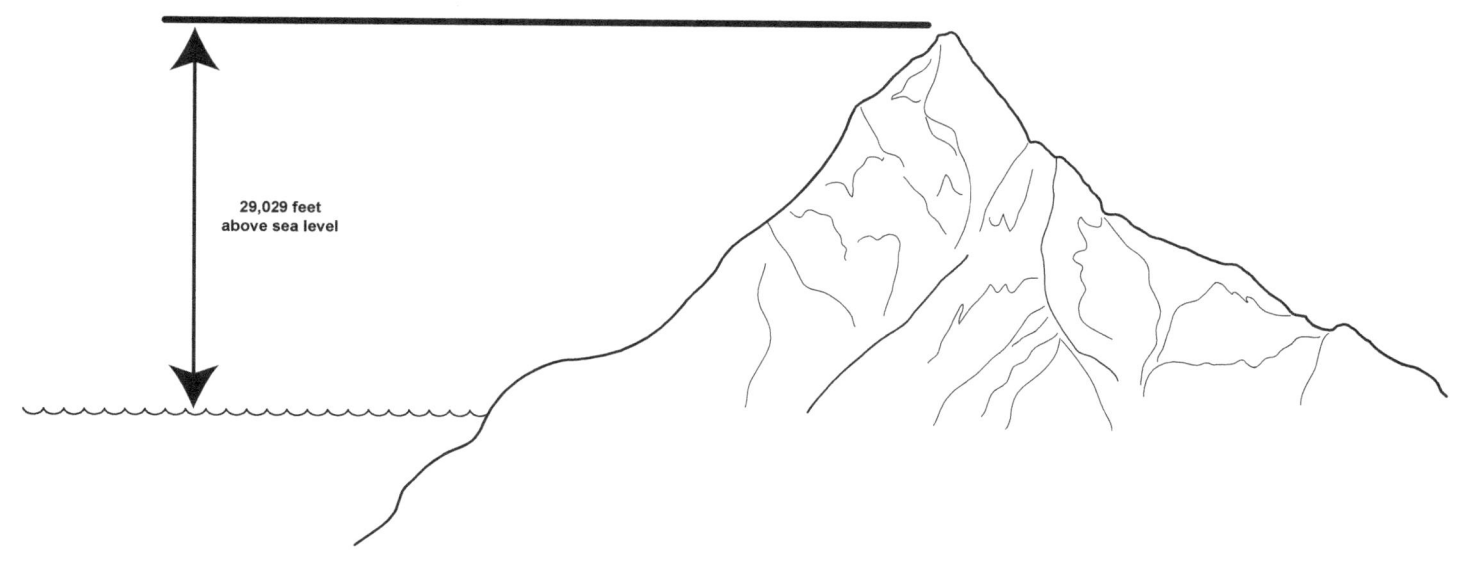

29,029 feet

◎ 1 2 3 4 5 6 7 8 9 10 11 12 13 14 15 16 17 18 19 20

MOUNT EVEREST IS 29,029 FEET TALL.

MOUNT EVEREST IS NOT NEXT TO THE OCEAN, BUT MOUNTAINS ARE MEASURED FROM THE OCEAN SURFACE TO THE TOP OF THE MOUNTAIN.

29,029 FEET = 8,848 METERS

The Little Engineer Coloring Book: Numbers

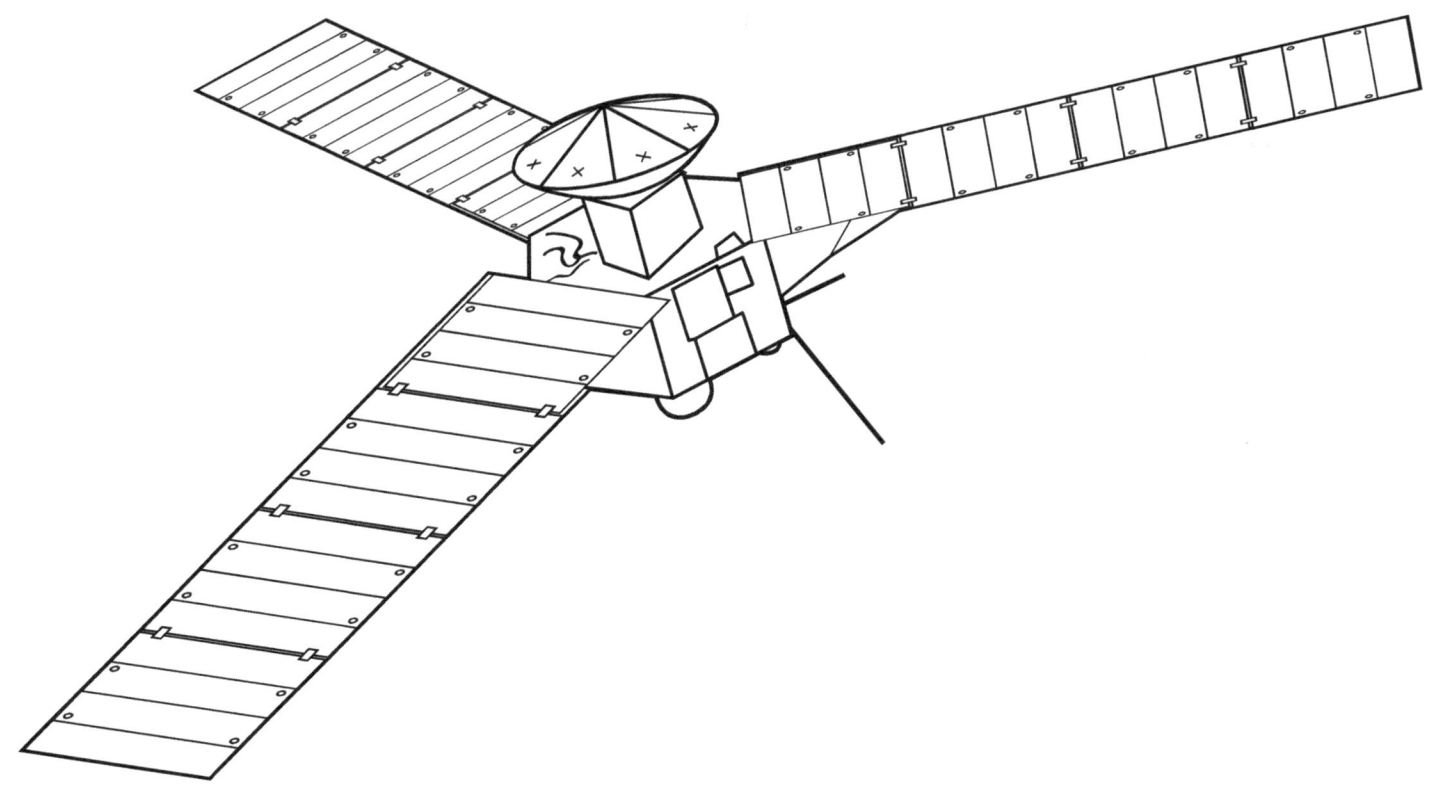

160,000 MPH

◎ 1 2 3 4 5 6 7 8 9 10 11 12 13 14 15 16 17 18 19 20

THE FASTEST SPACE PROBE WENT 160,000 MPH.

SPACE PROBES ARE SCIENCE EXPERIMENTS WE HAVE SENT OUT INTO SPACE TO STUDY OTHER PLANETS OR THE SUN.

160,000 MPH = 257,495 KM/H (KILOMETERS PER HOUR)

670,616,629 MPH

0 1 2 3 4 5 6 7 8 9 10 11 12 13 14 15 16 17 18 19 20

LIGHT TRAVELS AT 670,616,629 MPH.

WOW, THAT'S FAST! LIGHT IS INCREDIBLY FAST. WE ARE UNSURE IF ANYTHING TRAVELS FASTER THAN THE SPEED OF LIGHT.

670,616,629 MPH = 1,079,252,848 KM/H (KILOMETERS PER HOUR)

Congratulations! Training Complete

You are now certified and have successfully completed your House Building Training!

Go to: thelittleengineerbooks.com/tle-numbers-certificate or scan the QR code, and enter your email to get a ready-to-print certificate

Certificate of Completion

0 1 2 3 4 5 6 7 8 9 10 11 12 13 14 15 16 17 18 19 20

This Certifies That

_____

Has Successfully Trained & Completed the

NUMBERS TRAINING

And is Awarded This Certificate by

THE LITTLE ENGINEER

Parent/Guardian Signature: _____ Date: _____

0 1 2 3 4 5 6 7 8 9 10 11 12 13 14 15 16 17 18 19 20

SPECIAL PREVIEW

Check out this short preview of another fun coloring book!

LET'S GET STARTED!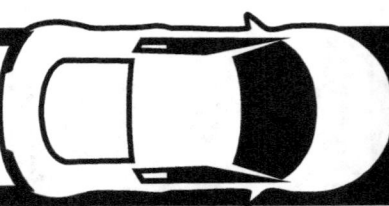

RADIATOR

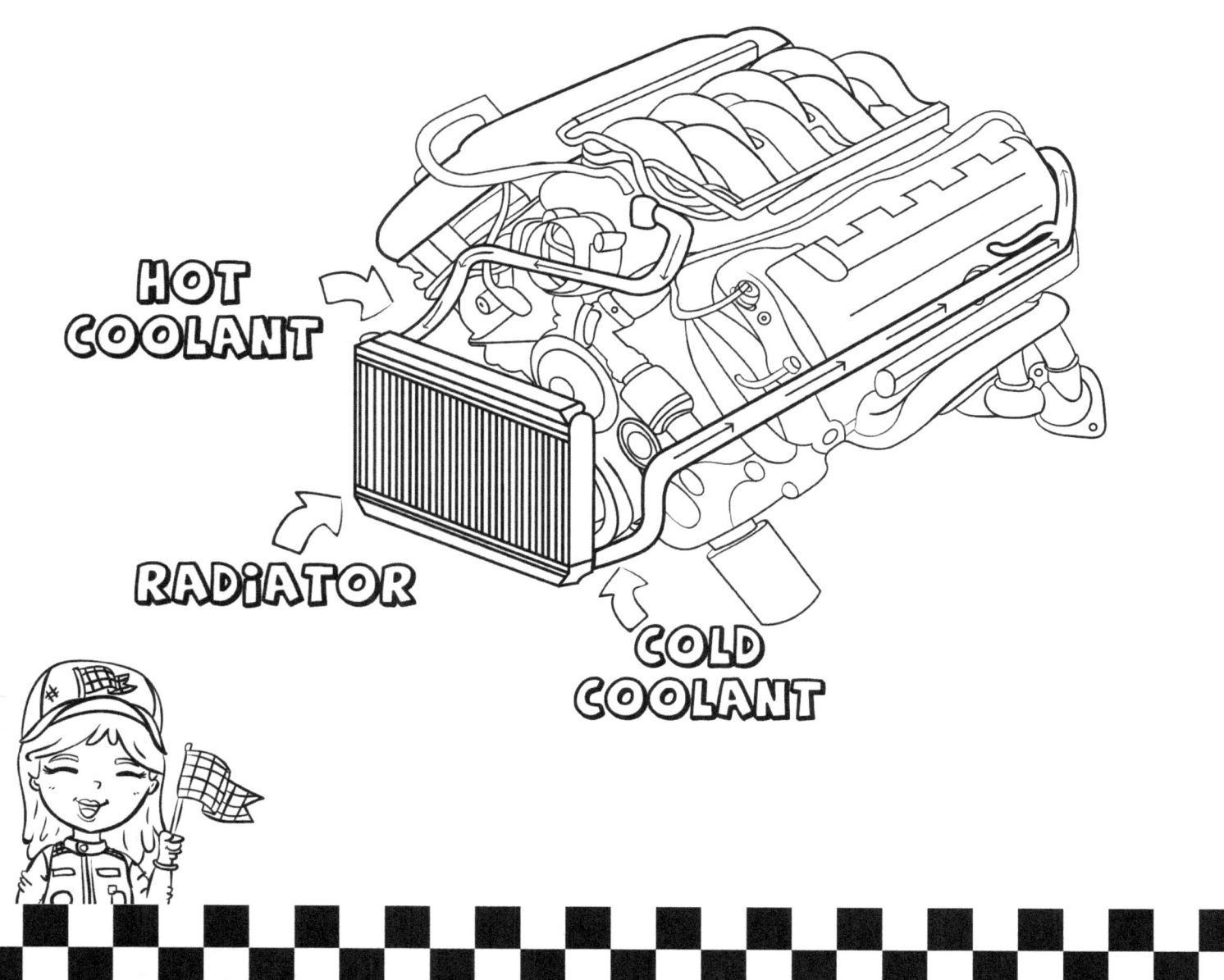

ENGINES GET VERY HOT.

- A RADIATOR HELPS KEEP THE ENGINE FROM GETTING TOO HOT.
- A WATER PUMP MOVES A SPECIAL LIQUID CALLED COOLANT THROUGH THE ENGINE AND THEN THROUGH A RADIATOR.

ENGINE ACCESSORIES

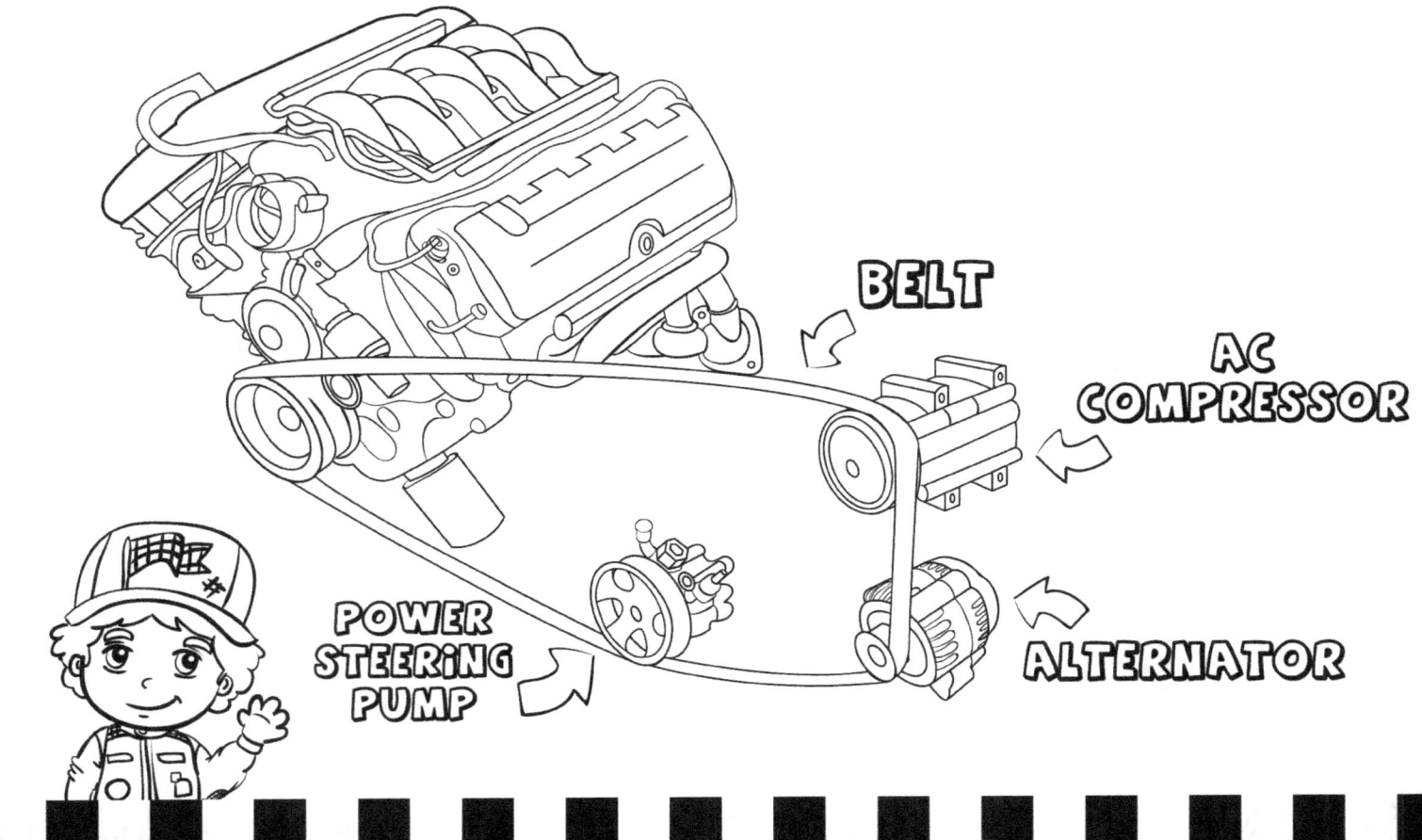

THE ENGINE ACCESSORIES ARE POWERED BY A BELT ON THE ENGINE. THE BELT SPINS THE WHEELS ON THE ACCESSORIES WHEN THE ENGINE IS ON.

- THE ALTERNATOR IS THE BATTERY CHARGER FOR THE CAR. IT IS A SMALL POWER GENERATOR THAT MAKES SURE YOUR CAR HAS PLENTY OF ELECTRICITY FOR LIGHTS, SPARK PLUGS, THE RADIO AND MORE.
- THE AC COMPRESSOR HELPS THE AC SYSTEM WORK SO THE AIR IS NICE AND COLD INSIDE THE CAR.
- THE POWER STEERING PUMP MAKES IT EASIER TO TURN THE STEERING WHEEL.

TWIN TURBOCHARGERS

THIS CAR HAS 2 TURBOCHARGERS!

MOST TURBOCHARGERS ARE UNDER THE HOOD AND HARD TO SEE, BUT SOME CARS HAVE THEM STICKING OUT OF THE HOOD WHICH IS REALLY COOL!

2 Differentials

ON TRUCKS, YOU CAN SOMETIMES SEE 2 DIFFERENTIALS. THIS MEANS THE TRUCK HAS 4-WHEEL DRIVE. SOME CARS HAVE 4-WHEEL DRIVE, BUT IT IS HARD TO SEE THE DIFFERENTIAL BECAUSE THE CAR IS CLOSE TO THE GROUND.

We hope you enjoyed the book!
Contact us anytime at CreativeIdeasPublishing.com

We are a US based publisher that consist of parents and teachers. We try our best to make products that our kids will love and we hope your kids love them too!

Ask your bookstore for more great titles from
Creative Ideas Publishing!

  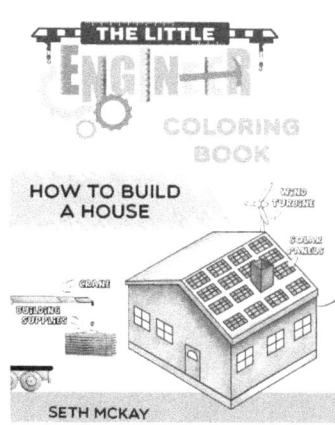 

 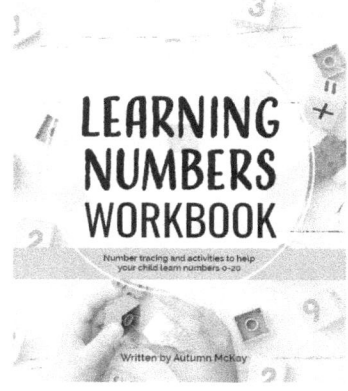